T0353446

Does Mathematical Study Develop Logical Thinking?

Testing the Theory of Formal Discipline

Does Mathematical Study Develop Logical Thinking?

Testing the Theory of Formal Discipline

Matthew Inglis

Nina Attridge

Loughborough Univesity, UK

World Scientific

NEW JERSEY · LONDON · SINGAPORE · BEIJING · SHANGHAI · HONG KONG · TAIPEI · CHENNAI · TOKYO

Published by

World Scientific Publishing Europe Ltd.

57 Shelton Street, Covent Garden, London WC2H 9HE

Head office: 5 Toh Tuck Link, Singapore 596224

USA office: 27 Warren Street, Suite 401-402, Hackensack, NJ 07601

Library of Congress Cataloging-in-Publication Data
Names: Inglis, Matthew. | Attridge, Nina.
Title: Does mathematical study develop logical thinking? : testing the theory of formal discipline /
    Matthew Inglis (Loughborough Univesity, UK) & Nina Attridge (Loughborough Univesity, UK).
Description: New Jersey : World Scientific, 2016. | Includes bibliographical references.
Identifiers: LCCN 2016013685 | ISBN 9781786340689 (hc : alk. paper)
Subjects: LCSH: Mathematics--Study and teaching. | Logic, Symbolic and mathematical. |
    Educational psychology. | Formal discipline.
Classification: LCC QA8.7 .I54 2016 | DDC 153.4/33--dc23
LC record available at https://lccn.loc.gov/2016013685

British Library Cataloguing-in-Publication Data
A catalogue record for this book is available from the British Library.

Desk Editors: V. Vishnu Mohan/Mary Simpson

Typeset by Stallion Press
Email: enquiries@stallionpress.com

Printed in Singapore

To Sam

Acknowledgements

We are extremely grateful to both the Royal Society and the Worshipful Company of Actuaries (WCA) for funding the research reported here, through a research fellowship to MI. Without their generous sponsorship we would not have been able to embark upon, let alone complete, this project.

We also need to thank those collaborators with whom aspects of the work reported here was conducted. In particular, important contributions were made by Lara Alcock, Toby Bailey, Pamela Docherty, Maria Doritou, Kristen Lew, Pablo Mejía-Ramos, Paolo Rago, Chris Sangwin, Adrian Simpson, David Wainwright, Elaine Wainwright, and Derrick Watson.

Although not directly involved in the research discussed here, our work would not have been possible without the collegiate research environment at Loughborough University's Mathematics Education Centre. We are grateful to all our colleagues — especially Lara Alcock, Sophie Batchelor, Camilla Gilmore and Ian Jones — for providing such consistently stimulating discussion.

Large portions of this book were written in two locations that both proved to be extremely productive working environments: the Bromley House Library in Nottingham and the Tryst Coffeehouse in Washington DC. Thanks to both.

Finally, we must thank all the research participants who gave up their time to make the work reported here possible.

Matthew Inglis and Nina Attridge

Contents

List of Figures

List of Tables

List of Abbreviations

A level Advanced level. An optional two-year qualification that is usually taken by 16–18 year olds after school but before university in the UK.

AC Affirmation of the consequent. An inference of the structure 'if p then q; q; therefore p'.

ANOVA Analysis of Variance. A statistical test for comparing means across groups.

ANCOVA Analysis of Covariance. A statistical test for comparing means across groups while controlling for a covariate.

AS level Advanced subsidiary level. The first year of an A level course.

CRT Cognitive Reflection Test. A measure of thinking disposition.

DA Denial of the antecedent. An inference of the structure 'if p then q; not p; therefore not q'.

DOI Decision Outcomes Inventory. A self-report measure of an individual's bad real-world outcomes.

INI Implicit Negation Index. A behavioural measure of the implicit negation effect.

MP Modus Ponens. An inference of the structure 'if p then q; p; therefore q'.

MT Modus Tollens. An inference of the structure 'if p then q; not q; therefore not p'.

NCI Negative conclusion index. A behavioural measure of the negative conclusion bias.

NFC Need for Cognition. A measure of thinking disposition.

TFD Theory of Formal Discipline. The theory which states that studying mathematics develops general thinking skills.

Chapter 1

The Theory of Formal Discipline

Why should students study mathematics? Over the last few years, more British 16 year olds have taken mathematics examinations than any other subject. Figure 1.1 paints a clear picture: only English gets close to having the same number of students as mathematics.[1] Students rate mathematics as being the most important school subject,[2] and in most countries they are required to study it for more hours per week and until an older age than other disciplines.[3] But what justifies the subject's privileged place in the eyes of students and on the school curriculum?

There have been many attempts to answer this question. They tend to fall into one of three categories: arguments based on the usefulness of mathematics, arguments based on its cultural importance, and arguments based on versions of the Theory of Formal Discipline (TFD), the main focus of this book. Before getting onto the TFD, let us briefly consider other attempts to justify place of mathematics in the curriculum.

Some suggest that mathematics must have a privileged status because it is of immense use to students in their later lives. These utilitarian arguments seem very persuasive, and indeed urgent. It has been found that around a quarter of adults in the UK struggle with basic mental arithmetic.[4] Such problems can have a serious impact upon an individual's quality of life: low levels of numeracy are associated with higher levels of unemployment, lower rates of pay, and higher levels of job insecurity while in employment.[5] While this seems to be a strong argument for insisting that all students develop basic number skills, some suggest that it does not justify insisting that students study advanced mathematical topics, such as geometry or algebra. John White, a philosopher of education, went as far as arguing that the

---

[1] Joint Council for Qualifications, 2014.
[2] Lamont & Maton, 2008.
[3] Hodgen *et al.*, 2010.
[4] Gross, 2008.
[5] Parsons & Bynner, 2005.

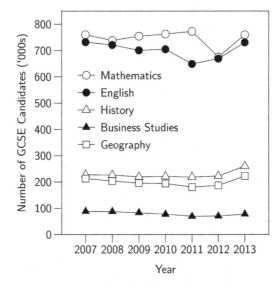

Figure 1.1   The number of candidates taking business studies, English, history and mathematics GCSEs, the qualification taken by 16-year-old school leavers in England and Wales.

only compulsory mathematical study which can be justified on utilitarian grounds is that which currently takes place in primary schools.[6] Although an appreciation of more advanced mathematics is demonstrably necessary for some scientific and commercial careers, White suggested that such topics should be studied voluntarily by those who are intrinsically interested in the subject, or who are minded to pursue a career that requires it. So although utilitarian arguments appear highly persuasive for lower-level mathematics, they have been challenged as a justification for the continued importance of mathematics in the curriculum up until the age of 16.

Others justify the importance of mathematics by appealing to its cultural and aesthetic properties. For many people, the certainty and precision of mathematics makes it a beautiful and elegant subject which enriches their lives; so it is natural for them to want others to experience the power of advanced mathematical thought for themselves. Mathematics is, according to this line of argument, an intrinsically valuable part of human culture and so should be studied by everyone. Fred Clarke, director of London's Institute of Education in the 1940s, put forward this position

---

[6]White, 2000.

in stark terms: "the ultimate reason for teaching long division to little Johnny is that he is an immortal soul."[7] But these appeals to mathematical aesthetics and cultural importance have also been challenged. For instance, some philosophers argue that there is no good reason to suppose that mathematics really has any genuinely aesthetic properties at all.[8]

The third common way of justifying mathematics' place in the school curriculum focuses on the cognitive effects of studying advanced mathematics. Mathematics should be studied because doing so develops general thinking skills that are useful through life. This position — which has been referred to as the Theory of Formal Discipline (TFD) — has a strong historical pedigree. Plato was an early advocate, and in *The Republic* wrote that

> Those who have a natural talent for calculation are generally quick at every other kind of knowledge; and even the dull, if they have had an arithmetical training, although they may derive no other advantage from it, always become much quicker than they would otherwise have been.[9]

This quote very clearly encapsulates the key tenets of the TFD. Plato was claiming not only that those who are mathematically able are also talented in many other domains, he also insists that this link is causal: that studying mathematics improves ones thinking more generally, and that therefore mathematical study is extremely important. From this, Plato drew policy implications arguing that future societal leaders should be persuaded to study mathematics. He also took a hard line on those who he deemed were not sufficiently mathematical, by having the phrase "let no one ignorant of geometry come under my roof" engraved above his front door.

Plato is not unusual among great thinkers for having endorsed the TFD. Numerous other examples can be cited. The important eighteenth century liberal philosopher John Locke, for instance, argued that mathematics should be studied by "all those who have time and opportunity, not so much to make them mathematicians as to make them reasonable creatures."[10] He went on to suggest that "having got the way of [mathematical] reasoning [...] they might be able to transfer it to other parts of knowledge as they shall have occasion."[11] In a similar vein, the prolific eighteenth century

---

[7] Cited by White, 2000, p. 72.
[8] Rota, 1997; Todd, 2008.
[9] Plato, 375BC/2003, p. 256.
[10] Locke, 1706/1971, p. 20.
[11] Locke, 1854, p. 493. Although Lester Mann points out that in some other pieces of writing, Locke appears to be more equivocal about the TFD (Mann, 1979).

hymn writer, Isaac Watts, suggested that

> If we pursue mathematical Speculations, they will inure us to attend closely to any Subject, to seek and gain clear Ideas, to distinguish Truth from Falsehood, to judge justly, and to argue strongly."[12]

Similarly, Francis Bacon suggested that mathematical study developed a man's wit and cunning,[13] and the English puritan Philip Doddridge believed that it teaches "attention of thought, and strength, and perspicuity of reasoning" and allows us "to distinguish our ideas with accuracy, and to dispose our arguments in a clear, concise and convincing manner."[14] It is straightforward to find many other examples of similar claims being made by important thinkers throughout history.

But these views are not simply constrained to philosophers and statesmen. They also influenced the design of the first school curricula. The historian Geoffrey Howson has documented the introduction of the school curriculum in England, and suggests that views consistent with the TFD were important drivers of placing mathematics at the heart of children's education.[15] In a series of lectures on teaching delivered in Cambridge in 1880, the year in which education became compulsory in England, J. G. Fitch remarked that:

> Our future lawyers, clergy, and statesmen are expected at the University to learn a good deal about curves, and angles, and numbers and proportions; not because these subjects have the smallest relation to the needs of their lives, but because in the very act of learning them, they are likely to acquire that habit of steadfast and accurate thinking, which is indispensable to success in all the pursuits of life. What mathematics therefore are expected to do for the advanced student at the University, Arithmetic, if taught demonstratively, is capable of doing for the children even of the humblest school.[16]

Interestingly, Fitch used the TFD to argue that it was important for girls to learn mathematics. He suggested that arithmetic had traditionally held a lesser place in girls' schools compared to boys' schools, because focusing on the utilitarian and practical value of mathematics was more common among school mistresses than it was among school masters. This, he argued,

---

[12] Watts, 1801, p. 113.
[13] Bacon, 1625.
[14] Cited by Howson, 1982, p. 53.
[15] Howson, 1982.
[16] Fitch, 1883, p. 320.

was a mistake:

> If, in short, the study of Arithmetic is mainly helpful in shewing what truth is, and by what methods it is attained, then surely it bears just as close a relation to the needs of a woman's life as to those of a man. For she, too, has intellectual problems to solve, books to read, and opinions to form.[17]

The TFD was also a curriculum orthodoxy in the United States throughout the nineteenth century, a period during which formal education became compulsory.[18] This influence continued well into the twentieth century. For instance, in the 1930s the influential mathematician Harold Fawcett designed a course in deductive geometry which had the explicit goal of preparing students to reason logically in day-to-day life:

> Fundamentally the end sought is for the student to acquire both a thorough understanding of certain aspects of logical proof and such related attitudes and abilities as will encourage him to apply this understanding to a variety of life situations.[19]

This was a widely held view in the US mathematics education community during the first half of the twentieth century. Indeed, it was explicitly endorsed by several presidents of the National Council of Teachers of Mathematics from the 1930s, including William Betz (President from 1932 to 1934), Halbert Christofferson (1938–1940) and Bruce Meserve (1964–1966).[20]

Two obvious questions emerge from this consideration of the views of historical thinkers and educators. While Plato, Locke, Bacon and others may have strongly endorsed the TFD, and while this may have influenced the design of nineteenth and early twentieth century school curricula, are there good reasons to suppose that the view is still important today? Does the TFD have any sway over how mathematics is viewed in modern times?

1.1 The TFD Today

In our view, the TFD is still an extremely important driver of educational policy, at least in the UK. In this section, we outline three sources of evidence which support this suggestion. First, we discuss a range of recent

---

[17]Fitch, 1883, p. 293.
[18]Stanic, 1986.
[19]Bennett *et al.*, 1938, p. 188.
[20]González & Herbst, 2006.

policy reports concerning mathematics education, each of which seems to endorse a version of the TFD. Second, we report a series of interviews we conducted with important stakeholders in the British mathematical community. Finally, we report a large-scale survey designed to investigate the extent to which recent mathematics graduates endorse the theory.

Over the last few years, a great many policy reports have been issued which have in one way or another attempted to influence the British government's policies concerning mathematics education. Indeed, according to one analysis over 50 such reports were issued during the two years from 2011.[21] Some were commissioned by government, and some were commissioned by the mathematical or scientific communities. Many contain sections which sought to defend the continued importance of mathematics on the school curriculum. Offering a few examples is sufficient to give a flavour of how the TFD is used in policy discussions.

The influential Cockcroft Report, published in 1982 suggested that, if taught appropriately, mathematical study could "develop powers of logical thinking, accuracy and spatial awareness".[22] Twenty years later, Adrian Smith, who was to become the Government's Chief Scientific Advisor, went further, and gave a particularly clear endorsement of the TFD:

> Mathematical training disciplines the mind, develops logical and critical reasoning, and develops analytical and problem-solving skills to a high degree.[23]

On the basis of this and other considerations, Smith recommended that mathematics students should enjoy tuition fee rebates, and that teachers should be financially rewarded for attending mathematics professional development courses. Just prior to the 2010 UK general election, the TV personality Carol Vorderman was commissioned by the British Conservative Party to write a report on how mathematics education could be improved. She endorsed Smith's view, suggesting that mathematics should be compulsory until the age of 18, in part because

> Mathematics is not only a language and a subject in itself, but it is also critical in fostering logical and rigorous thinking: as such its influence is immense.[24]

---

[21] http://mathsreports.wordpress.com/2013/01/05/homehome/
[22] Cockcroft, 1982, p. 2.
[23] Smith, 2004, p. 11.
[24] Vorderman *et al.*, 2011, p. 3.

These examples, and others like them, suggest that the TFD is still influencing policy discussions today. But is this correct? When pushed, do those in positions of influence in the mathematical community support the TFD?

Soon after the publication of the Vorderman Report we, along with our colleagues Elaine and David Wainwright, investigated this question by interviewing eight influential figures in the mathematics education community about their views on the TFD.[25] The interviewees had varied backgrounds, shown in Table 1.1.[26] Many were affiliated to one or more of the subject associations that promote mathematics in the UK. For instance, some were involved in developing the education policies of the Institute of Mathematics and its Applications or the London Mathematics Society (the learned societies which deal with applied and pure mathematics, respectively). One was a Member of Parliament who had contributed

Table 1.1   Participants in the interview study.

Participant	Role
Participant 1	Academic mathematician, textbook writer. Member of the Higher Education Academy Mathematics Subject Centre.
Participant 2	Academic mathematician. Member of the Council of the Institute of Mathematics and its Applications.
Participant 3	Academic mathematician, textbook writer. Member of the Education Committees of the Institute of Mathematics and its Applications and the London Mathematical Society.
Participant 5	Academic mathematician. Member of the Higher Education Academy Mathematics Subject Centre.
Participant 6	Academic mathematician. Senior manager at a teaching focused university, member of the Higher Education Academy Mathematics Subject Centre.
Participant 7	UK Member of Parliament with an interest in education policy.
Participant 8	Academic mathematician. Senior manager at a research intensive university, member of the Council of the Institute of Mathematics and its Applications.
Participant 9	Academic mathematics educator. Member of the Council of the Institute of Mathematics and its Applications, contributor to several influential reports on mathematics education.

[25]Wainwright, Attridge, Wainwright, Alcock, & Inglis, 2015.
[26]Nine participants agreed to be interviewed but participant four withdrew before the interview was conducted.

to several important education policy developments, and several were affiliated with the Higher Education Academy's Mathematics Subject Centre, a body designed to promote high-quality mathematics teaching at the undergraduate level.

We started the interviews by presenting participants with a series of quotes supporting the TFD, including those from Plato and the Smith Report cited earlier, and asking them for their reactions. All eight of the interviewees gave several clear endorsements of the TFD. Participant 5, for instance, said "I believe that that ability to learn, to control and manipulate abstract ideas in a logical and analytical way [. . .] is a process you do get better at as you do more mathematics", and Participant 1 suggested that studying mathematics "probably does give you some structure to the way you think about things than perhaps some other people that haven't had some sort of mathematical training [. . .] And without any doubt I do believe that mathematical training gives you a process for going about problem solving".

But alongside widespread clear endorsements were some more nuanced responses. When asked directly whether studying advanced mathematics developed general thinking skills, Participant 5 responded by saying "It's a very important question. It makes one a bit nervous in case the answer is no", and that if the answer did turn out to be no, then "I think we might well suppress [the] evidence by ignoring it".

Some of our interviewees suggested that if mathematics were taught appropriately it would develop the kinds of skills mentioned in the Smith Report — logical reasoning and problem solving — but that if teaching were in some way suboptimal it might not. Participant 8 remarked that "I think it's possible to pass a maths degree, particularly a joint honours maths degree, without developing many of these [skills] to a particularly high level", and when shown a specific reasoning task (the conditional inference task, discussed in detail later) Participant 6 said "if maths graduates aren't doing better on [this task] than non-maths graduates then I yeah I think we are doing something wrong".

Several interviewees raised the possibility of what we call the *filtering hypothesis*: the idea that instead of mathematical study developing reasoning skills, those who are already better reasoners are more likely to be filtered into studying post-compulsory mathematics. Participant 2 referred to this as the 'chicken and egg' situation: "What I'm not clear about here is the chicken and egg situation, so I'm not clear whether or not people that go into maths are already logical". Participant 5 agreed, saying that "The problem I have with all of this is it's a kind of philosophical problem underlying all of these statements is that people who gravitate towards mathematics, mathematical studies, who are enthralled by mathematical ideas, and even those who like calculation, they're a bit of a self-selecting bunch ... And it's not always clear whether these people are bright because they are generally bright and they like maths or calculation or because they've done a lot of it and their minds are improved". An important goal of several of the studies discussed later in the book is to distinguish between the TFD and this alternative filtering hypothesis.

When asked why they tended to support the TFD, our interviewees mainly referred to two kinds of evidence. The first was personal experience. Participant 7, a Member of Parliament who has strongly influenced UK education policy, exemplified this by simply stating "I am quite a firm supporter of [the TFD] really ... it's just my own experience of studying the subject". Participant 8, a senior manager at a research-intensive university, reported that his academic background sometimes became obvious when engaging with non-mathematicians: "you know when I'm sitting around with other senior managers most of whom, or all of whom, are not mathematicians and some of them are from completely different parts of the academic spectrum, and people will say to me when I say something, oh that's the mathematician in you, you know because you think through things in a different way."

A second justification given by our interviewees for their support of the TFD was the continued success of mathematics graduates at finding employment. Participant 8 justified his stance by saying "I'm kind of relying on just the fact that [employers] keep employing [mathematics graduates] as being evidence that they are useful but I would imagine that if I was an employer and I needed people who were going to be logical and critical and analytical and be able to solve problems and I saw somebody who had a good class maths degree I think they would be a good bet". And Participant 9 agreed, arguing "undoubtedly, in practice, out there [mathematics is] used as a way of sorting out people and these are very hardnosed finance people and business people ... and they use mathematics as a sieve. Now, they might not know why but it presumably has worked quite well."

Is it the case that mathematics graduates obtain more or better jobs than their non-mathematical colleagues? The UK's Higher Education Statistics Agency collects data on such matters. Figure 1.2 shows the percentage of UK graduates from three different disciplines who were working or studying six months after they graduated. On this measure

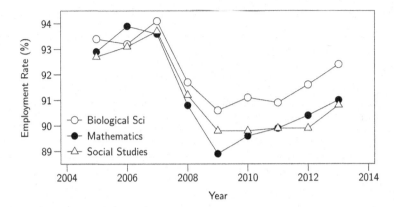

Figure 1.2   The employment rate of UK graduates from three disciplines, six months after graduation (*Source*: Higher Education Statistics Agency).

mathematics graduates do not appear to outperform other graduates. However, it may be the case that mathematics graduates tend to earn more than those from many other disciplines. For instance, in 2012 the median salary of UK mathematics graduates from the 2008/9 graduating cohort was £32,000, compared to £23,000 for biological scientists, £27,000 for social studies graduates, and the overall median of £26,500. In fact, according to these figures, only graduates in medicine and dentistry had higher median salaries than mathematicians.[27]

There is also evidence that earlier mathematical study leads to a salary premium in employment, even when attempts are made to control for possible confounding variables such as general academic achievement. In 2002, the economists Peter Dolton and Anna Vignoles published an analysis which suggested that students who opted to study A Level mathematics, the optional course in mathematics taken by 16–18 year olds in England and Wales, earned between 7% and 10% more than those who chose not to study mathematics. This advantage remained even after controlling for a variety of background variables, including initial academic ability.[28] A similar result was found when Michael Adkins and Andrew Noyes replicated this analysis using more recent data.[29] In sum, it does appear to be the case that studying mathematics is associated with higher salaries, but perhaps not with a lower chance of unemployment.

---

[27]Destinations of Leavers from Higher Education Longitudinal Survey, 2008/9.
[28]Dolton & Vignoles, 2002.
[29]Adkins & Noyes, 2016.

Overall, we found strong support for the TFD among our interviewees, all of whom are influential figures in the mathematical community. But, despite this headline support, consistent with the views of philosophers and educational report writers discussed earlier, there was a nuanced recognition that an important alternative hypothesis exists. This alternative, which we call the filtering hypothesis, suggests that the reason we associate the study of mathematics with the development of reasoning skills is simply that those who are talented reasoners are more likely to opt to study post-compulsory mathematics.

While many mathematicians and policymakers may endorse the TFD, can the same be said of students themselves? Is it the case that mathematics graduates, once they have entered the world of work, believe that their mathematical studies developed general reasoning skills? In 2011 we, along with our colleagues Tony Croft and Janette Matthews, conducted a survey of UK mathematics graduates to try to find out.[30]

We contacted graduates from the 2008/9 graduating cohort, obtaining 428 responses. This sampling strategy meant that all our respondents were around two and a half years into the next stage of their careers, so were well placed to reflect on their courses and employment to date. An analysis of responses suggested that around 46% of those who received our invitation to participate did so, indicating that our sample can be considered reasonably representative.

We asked the graduates a number of different questions concerning their views on the TFD. First, we asked them to state the extent to which their course assisted them in developing skills, using a four point Likert scale ('a lot', 'quite a bit', 'not much', 'not at all'). Second, we asked them to say how important these skills had been in their career so far, again with a four point Likert scale ('very important', 'quite important', 'not very important', 'not at all important'). Third, we explicitly presented them with Smith's version of the TFD ("Mathematical training disciplines the mind, develops logical and critical reasoning, and develops analytical and problem-solving skills to a high degree") and asked them to state the extent to which they agreed. Finally, we allowed them the opportunity to write any further comments that they had.

Table 1.2 shows the headline results from the survey. There was overwhelming agreement about both the importance of logical thinking, problem-solving and analytical skills to the graduates' careers, and that

---

[30]Inglis, Croft, & Matthews, 2011.

Table 1.2   The percentage of mathematics graduates in the survey who thought that (a) the given skill was 'very' or 'quite' important for their career and (b) the given skill was developed 'a lot' or 'quite a bit' during their undergraduate mathematics course.

Skill	Agree Important (%)	Agree Developed (%)
Logical thinking	97	96
Analytical approach to working	96	94
Problem-solving skills	95	94

their mathematical studies had helped them develop these skills. When presented explicitly with Smith's version of the TFD, there was again strong support for the theory. Overall 81% agreed with Smith's formulation of the theory, and 44% strongly agreed. As one might expect from these figures, support for the TFD was also found in the free-text comments. For example, one graduate wrote that it was the rigorous subject matter that had "developed [his/her] most important skill: problem-solving, analytical thinking and the tenacity to keep trying until a solution is found."

In sum, as with our interviewees, we found strong support for the TFD among recent mathematics graduates. Almost all believed that their mathematical studies had developed skills of problem solving, logical thinking and their ability to adopt an analytical approach; and a large majority endorsed Smith's formulation of the TFD, and accepted that "mathematical training disciplines the mind, develops logical and critical reasoning, and develops analytical and problem-solving skills to a high degree."

So far we have discussed four sources of support for the TFD. First, many great thinkers throughout history — including influential philosophers such as Plato and Locke — have endorsed the theory. Second, modern policymakers, at least in the UK, have often cited versions of the theory when attempting to influence government education policy. Third, we have reported interviews with eight leading stakeholders in the mathematical community.[31] All endorsed the TFD, albeit with caveats in some cases. Finally, we have shown that the theory enjoys near universal support among mathematics graduates. Overwhelming majorities of graduates believe that their mathematical studies developed the generic thinking skills that are the focus of the TFD, and that these skills have been useful to them in the workplace.

---

[31]Wainwright *et al.*, 2015.

It is very noticeable that none of these sources of support for the TFD include any direct empirical evidence of mathematical study being associated with improvements in thinking skills. Largely, they are based on individuals reflecting on their own experiences of learning mathematics. And, as several of our interviewees explicitly noted, relying upon the personal views of those who are enthusiastic about mathematics and mathematics education may result in biased assessments of the validity of the TFD. An obvious question emerges: is there any empirical research which justifies the strong faith in the TFD shown by our great thinkers, our policymakers, our stakeholders and our mathematics graduates? This is the topic to which we turn next.

1.2   Transfer and the Educational Psychologists

Classically, the TFD was applied to many more subjects than just mathematics. For instance, the school teacher John Tarver, writing at the turn of the nineteenth century, argued that children must learn Latin:

> My claim for Latin, as an Englishman and a [teacher], is simply that it would be impossible to devise for English boys a better teaching instrument. [...] The acquisition of a language is educationally of no importance; what is important is the process of acquiring it. [...] The one great merit of Latin as a teaching instrument is its tremendous difficulty.[32]

The American founding father and philosopher Benjamin Franklin made similar claims about chess, in an essay published in 1786. He wrote that

> The game of Chess is not merely an idle amusement. Several very valuable qualities of the mind, useful in the course of human life, are to be acquired or strengthened by it, so as to become habits, ready on all occasions.[33]

Tarver and Franklin's arguments are directly analogous to that made by Locke when he suggested that mathematical study was important, not because it would teach you mathematics, but because it made you a "reasonable creature". But few subscribe to the TFD today in the context of Latin or chess. Why not? The answer is that during the course of the

---

[32] Quoted by Monroe, 1909, p. 511.
[33] Quoted by Richardson, 1979, p. 59.

twentieth century a large body of opinion has built up in the educational psychology community which suggested that the TFD is simply false.[34]

The first serious attempt to challenge the TFD came from Edward Thorndike, one of the first educational psychologists, who spent his career at Columbia University's Teachers College. Born in 1874, Thorndike's early work was on animal learning; he studied under William James at Harvard and James Cattell at Columbia, and pioneered the study of animal intelligence.[35] His work on animal learning led him to formulate the influential 'Law of Effect', a precursor of the behaviourist movement in psychology. This law stated that if an animal's action is followed by a "satisfying state of affairs" in a given situation, then it is more likely to occur in that situation in the future. Equally, if an action is followed by an "annoying state of affairs" it is less likely to occur again. In other words, responses which lead to positive outcomes are more likely to occur again, and responses which lead to negative outcomes are less likely to.

Thorndike also made substantial contributions to statistical methods in psychology, to the nature-nurture debate, and to the development of educational and intelligence tests. He firmly believed that introspective accounts, such as those given by Plato and Locke in the context of the TFD, should be rejected, and instead that the methods of science should be brought to bear on debates which had previously remained outside the domain of empirical data. He wrote that "anything that exists, exists in some amount and can be measured", and that therefore it ought to be.[36]

For Thorndike, the TFD was of much wider interest than just mathematics. He described the issue in these terms:

> The problem of how far the particular responses made day by day by pupils improve their mental powers in general is called the problem of the disciplinary value or disciplinary effect of studies, or more briefly, the problem of formal discipline. How far, for instance, does learning to be accurate with numbers make one more accurate in keeping his accounts, in weighing and measuring, in telling anecdotes, in judging the character of his friends? How far does learning to reason out rather than guess at or learn by heart a problem in geometry make one more thoughtful and logical in following political arguments or in choosing a religious creed or in deciding whether it is best for him to get married? How far does the

---

[34] Mann, 1979.
[35] Gates, 1949.
[36] Quoted in Gates, 1949, p. 242.

habit of obedience to a teacher in school generate the habit of obedience to parents, laws and the voice of conscience?[37]

In *The Principles of Teaching*, published in 1906, Thorndike set out what he called 'the common view', a version of the TFD which suggested that every time a person acquired a new specialist ability it would contribute to a general ability: "Improved attention to grammar or Latin would thus mean an improvement of the power to attend to any task."[38] This common view could, said Thorndike, be shown to be false. He argued that the only way of testing the TFD would be to determine the extent to which training on some specialised skill produced improved performances in other domains.

In one classic experiment, Thorndike and his collaborator Robert Woodworth asked a group of adults to estimate the length of a series of long lines. Having taken this benchmark measure, they then trained their participants to estimate the length of short lines. They were given a short line, asked to estimate its length, then told the correct length and told to keep a record of their error. Once their performance on these short line tasks had reached a satisfactory level, the participants were given the long line task again. Thorndike and Woodworth found that their participants had made little or no improvement.[39]

Similar results were found in other domains: skills at estimating the area of rectangles apparently did not help much with estimating the area of circles or triangles.[40] Developing the ability to quickly mark verbs in pages of text did not develop the ability to quickly mark other types of words.[41] Students who were encouraged to produce neat schoolwork during their arithmetic classes, showed no such improvement in their language or spelling classes.[42]

From all these results, Thorndike concluded that the TFD was "much exaggerated" and that it should be abandoned. Instead he suggested that transferring skills from one domain to another, even if closely related, was impossible. If this were correct, the justification for teaching a school subject such as Latin or mathematics could not be based on claims about how it

---

[37]Thorndike, 1906, pp. 235–236.
[38]Thorndike, 1906, p. 236.
[39]Thorndike & Woodworth, 1901a; Thorndike, 1906.
[40]Thorndike & Woodworth, 1901c.
[41]Thorndike & Woodworth, 1901b.
[42]Bagley, 1905.

develops general reasoning. Thorndike wrote

> It is extremely unsafe to teach anything simply because of its supposed
> strengthening of attention or memory or reasoning ability or any other
> mental power; when a teacher can give no other reason for a certain
> lesson or method of teaching than its value as discipline, the lesson or
> method should be changed.[43]

But what about the effects of an academic education? Could it be that
the value of studying subjects such as mathematics or Latin cannot be
appropriately investigated by studying the value of learning to estimate the
length of short lines? Later in his career Thorndike began to study the issue
more directly. He first needed to develop a method of measuring the extent
to which students reasoned effectively. In the early 1920s, he published a
series of questions which, he hoped, did just that.

Thorndike's method involved asking a variety of different questions,
which covered four broad areas that he labelled selective thinking, relational
thinking, generalisation and organisation. He referred to the outcome of his
test as a measure of 'general intelligence', and drew heavily on existing
intelligence measures for his items. In one part of the test, he would list a
series of statements and ask his participants to say which could not possibly
be true. For example, which of the following statements are impossible?

(1) Each singer shouted at the top of his voice, but the big fat man with
    the red necktie could be heard above all others.
(2) The poor wanderer, finding himself without means of lighting his camp
    fire, made a fruitless search through his equipment by the light of a
    single candle.
(3) By the light of a dim lantern, the farmer found the source of the
    nauseating odour.[44]

Students would also be tested on their vocabulary. For instance, they might
be told to find words beginning with 'b' that were opposite to 'good',
'white', and 'spiritual'.[45] They might also be told to find "the word that
does not belong" in lists such as: stupidity, dullness, foolishness, dishonesty,
and ignorance.[46] Importantly, Thorndike also included several tests of
arithmetic and geometry. He asked his students to solve problems such

---

[43] Thorndike, 1906, pp. 242–243.
[44] The poor wanderer, of course, could have used his candle to light his camp fire.
[45] Bad, black, and blasphemous.
[46] Thorndike, 1922.

as "4 percent of $600 equals 6 percent of what amount?" and to rearrange the symbols "3 3 4 10 $= + -$" so that they made a true equation.

Thorndike used his newly developed test to investigate directly the extent to which studying different subjects improved students' general thinking skills. His design was straightforward. He split his test into two halves of roughly equal difficulty and, in May 1922, gave it to a large number of students from grades 9, 10, and 11. Half were given the first half of the test, and the remainder the second half. A year later he asked the same group to take the half they had not been given the year before. He also recorded which school subjects his participants had studied in the intervening period. The idea was that if there were systematic differences between the gain scores of students who had taken different subjects, that would provide strong evidence that studying the subjects had differing effects.[47]

The results were mixed. Overall, the extent to which students improved on the tests was only very weakly related to the subjects they had studied during the year. However, it was clearly the case that those who had higher scores in Year 1 gained more than those who had lower scores. This kind of finding, which can be colloquially characterised as 'the rich get richer and the poor get poorer', is sometimes referred to as the Matthew Effect, after a verse from the Gospel of Matthew ("For unto every one that hath shall be given, and he shall have abundance: but from him that hath not shall be taken even that which he hath." Matthew 25:29).

The Matthew Effect is an important threat to the validity of studies which aim to investigate the TFD. If two groups differ at Time 1 and also differ in the extent to which they improve from Time 1 to Time 2, then the Matthew Effect suggests that we cannot reliably distinguish between the TFD and its rival, the filtering hypothesis. Perhaps the difference in gain between Times 1 and 2 is simply due to the pre-existing differences at Time 1, coupled with the Matthew Effect.

Thorndike concluded his study by suggesting that there was little difference between the extent to which two students, with equal scores at Time 1, would develop if one took Latin and geometry and the other took cooking and sewing. But, he conceded that his findings were not conclusive, so a few years later set about replicating them. Along with his collaborators Cecil Broyler and Ella Woodyard, he found another large group of students

---

[47]Thorndike, 1924.

and tested them, following the same design, at the start and end of the academic year 1924/25.[48]

Thorndike again found that the differences between the subjects were small. However, those differences which did exist were not completely inconsistent with the TFD in the context of mathematics. Students who studied algebra, geometry or trigonometry showed the second highest gains in the study, once differences in incoming test scores were taken into account (an attempt to control for the Matthew Effect). However, the findings were less promising for Latin: after his statistical corrections, students who studied Latin actually showed a slight decrease in their scores.

Delving deeper into his data, Thorndike found that the academic subjects which appeared to show the highest gains did so unevenly across the different types of items in his test. So the gains showed by those studying algebra, geometry and trigonometry were very different from the gains showed by those studying Latin or French. Perhaps unsurprisingly, the group who studied mathematical subjects showed greater gains on the parts of the test concerned with numerical operations, and those studying languages showed greater gains on those parts of the test concerned with vocabulary. Thorndike concluded that his findings were consistent with his earlier studies on the difficulties with transferring abilities between related domains, writing:

> The superior gains which we have found associated with taking certain studies are then surely not to be regarded as consisting entirely in a general improvement for thinking with all sorts of data.[49]

Thorndike's conclusions, and particularly his Law of Effect, greatly contributed to the growing influence of behaviourism in psychology. This school of thought, perhaps most closely identified with the work of B. F. Skinner and John Watson, took the view that thought simply consists of responses to concrete stimuli, and that learning is merely the result of reinforced stimulus-response links. Clearly if learning is merely the consequence of responding to particular stimuli, there is no room for skills to be transferred from one context to another. While a student might be able to learn to reason logically in response to algebraic stimuli, in the presence of different, non-mathematical stimuli, this response would not be activated as the required stimulus-response link was different.

---

[48] Broyler, Thorndike, & Woodyard, 1927.
[49] Broyler *et al.*, 1927, p. 403.

While hugely impressive, even when evaluated by modern research standards, Thorndike's work on transfer and the TFD can be criticised on at least two grounds. First, some have suggested that Thorndike's work did not take sufficient account of the quality of his participants' learning. Perhaps they simply had not been taught effectively. Charles Hubbard Judd, a contemporary of Thorndike's, investigated this suggestion empirically.[50] He formed two groups of children, and explained the principle of the refraction of light to one group but not the other. Following this he asked both groups to practice throwing darts at targets submerged in 12 inches of water, finding no initial difference in performance between the groups. After this practice session, he assessed the performance of the groups when throwing at a target submerged by only 4 inches of water. Judd found evidence of transfer: the group who had learnt the underlying principle that applied to both settings performed better in the novel setting. Perhaps then, transfer is possible when learners extract appropriate general principles from a particular context.

In passing it is worth noting that Judd's 1908 report omitted many details of his experimental setup, including how he had managed to create a dart that could be accurately thrown into water. Indeed in the 1940s a replication attempt by Gordon Hendrickson and William Schroeder[51] determined that this was a highly non-trivial challenge. They found "a strong tendency for the dart to ricochet from the surface of the water" and that consequently "there was no certainty that the dart would travel in the direction desired." Nevertheless, when they replaced Judd's dart with an air gun and conducted an analogous study, they found similar results: those who had learnt about refraction scored better when shooting at a new target than those who had not.

A second major complaint about Thorndike's work concerns the test items in his 1920s studies. Although Thorndike found that those students who had studied mathematics showed noticeably higher gains on his measure of 'general intelligence', he attributed this to them merely scoring higher on the mathematics component of his measure. Naturally, this begs the question of why he chose to include items on the test which clearly related to the subject matter studied by some of his participants. This raises the more general issue of how to best measure reasoning behaviour. Since Thorndike's early work there has been a great deal of work in the

---

[50] Judd, 1908.
[51] Hendrickson & Schroeder, 1941, p. 206.

psychology of reasoning. Happily then, modern research on the same topic can draw on a substantial literature on reasoning measures, and we discuss ways of assessing students' reasoning performance in detail in the next chapter.

Notwithstanding these criticisms, Thorndike's work was highly influential and the belief that there were no general thinking skills of the type required by the TFD was the default position of most twentieth century psychologists. Even when the behaviourist paradigm, which had developed out of Thorndike's early work, was overthrown during the cognitive revolution in the 1960s, a belief in the domain specificity of thinking skills remained. Whereas the behaviourists believed that thinking could only be studied as stimulus-response links, this position was rejected by cognitive scientists who felt that carefully designed experiments could reveal internal mental processes. Nevertheless, Alan Newell summed up the standard cognitive anti-TFD view when he wrote that "the modern (i.e. experimental psychology) position is that learned problem-solving skills are in general idiosyncratic to the task."[52]

This line of behaviourist and cognitive research represents a serious challenge to the TFD. If there is no such thing as a domain-general thinking skill, then clearly the study of mathematics cannot develop any such skill.

Notwithstanding this general agreement, there was one influential mid-century psychologist who rejected the consensus that all thinking skills are domain specific. This led to another series of studies that made an important contribution to understanding the TFD. Unlike most of his peers, the Swiss developmental psychologist Jean Piaget *did* believe that it was possible to meaningfully talk about domain-general reasoning skills. In fact he thought that teenagers and adults reasoned in accordance with the norms of formal mathematical logic, writing that, by about the age of 12, "reasoning is nothing more than the propositional calculus itself."[53] However, Piaget's position was quite different to the TFD. Instead of claiming that studying formal subjects such as mathematics developed reasoning skills, he believed that the process of becoming a formal reasoner was a natural part of maturation. No educational intervention could speed the process up, and children would just have to wait until they reached the right age before developing these domain-general skills.

---

[52]Newell, 1980, p. 178.
[53]Inhelder & Piaget, 1958, p. 305.

Piaget's view has been comprehensively shown to be wrong. Adults often do not reason in accordance with formal logic, and certainly 12-year-olds do not. One of the earliest demonstrations of this came from the work of Peter Wason, an important early figure during the emergence of the psychology of reasoning as a field of study in the 1960s.

Wason, a psychologist from University College London, had a knack for developing reasoning tasks that offered surprising and important insights into how humans think. For instance, he developed an experimental test of Karl Popper's suggestion that falsifiability is critical to scientific hypothesis testing. Wason gave his participants the sequence '2 4 6' and told them that the numbers conformed to a simple rule, which they must discover by generating triples of their own. After presenting each triple, together with their rationale for choosing it, Wason told the participant whether or not it satisfied the rule.

An important finding emerged. Participants were much more likely to try to confirm their hypotheses than disconfirm them. If they thought the rule was "three successive even numbers", they would generate many instances of three successive even numbers, and not, as Popper would have advocated, test disconfirming instances. Few successfully identified the rule Wason had in mind (the rule merely stated that the numbers must be any three integers ordered by size).[54]

The task that Wason became most famous for has spawned an enormous literature (which we add to later in the book). It focused on how people reason about so-called conditionals, statements of the form 'if p then q'. The set up is simple. An experimenter shows the participant four cards, bearing the symbols D, K, 3, and 7. It is known that each card has a letter on one side and a number on the other. The task is to choose just those cards which you need to turn over in order to determine whether the following rule is true or false: "if the letter is D, then the number is 3".[55] The task, which became known as Wason's selection task, is shown in Figure 1.3.

Wason found a result quite inconsistent with Piaget's view that adults always reason in accordance with logical rules. He discovered that the most common response was for participants to answer that it was necessary to turn over the D and 3 card. But this is not consistent with norms of formal logic: given the rule 'if D then 3', the 3 card is irrelevant. If there is a D on its back, then it's consistent with the rule, and if there isn't then it's also

---

[54] Wason, 1960.
[55] Wason, 1966.

Four cards are placed on a table in front of you. Each card has a letter on one side and a number on the other. You can see:

| D | K | 3 | 7 |

Here is a conjectured rule about the cards:

"if a card has a D on one side, then it has a 3 on the other".

Your task is to select all those cards, but only those cards, which you would need to turn over in order to find out whether the rule is true or false.

Figure 1.3   The Wason selection task.

consistent because the rule is only about cards with Ds on. However, the 7 card *is* important: if it has a D on its reverse, then the rule is falsified. Overall, contrary to Piaget's expectations, less than 10% of educated adults made the normatively correct selection of D and 7. The infrequency with which people select D and 7 on this task is perhaps one of the most robust findings in the cognitive psychology literature.[56]

The selection task has generated an enormous literature, and quite why it is so difficult (and even whether the normatively correct answer should be considered to be the 'right' answer at all) has been hotly debated. For instance, some have argued that problems such as the selection task are merely 'cognitive illusions' that have no relevance to the real world.[57] Others have suggested that while the answer seen as correct by Wason is consistent with logical norms, it is non-adaptive in the sense that the more common D3 selection will yield more information, given certain sensible assumptions.[58] We discuss some of these issues in more detail later in the book. For now, however, we report just one other important selection task finding, which is relevant to the issue of transfer of reasoning skills.

A few years after introducing the selection task, Wason tried a variant where he replaced the abstract letters and numbers with real-world content. In this version, the rule became "every time I go to Manchester I travel by train" and the cards 'Manchester', 'Leeds', 'Train' and 'Car'.[59] Surprisingly, the majority of participants now made the correct selections: Manchester

---

[56] Wason & Johnson-Laird, 1972.
[57] Cohen, 1981; Lopes, 1991.
[58] Oaksford & Chater, 1994.
[59] Wason & Shapiro, 1971.

and Car. Apparently, the way in which adults reason depended on the context in which the reasoning took place. Even structurally identical problems apparently elicited different responses. This was another serious problem for the notion of transfer and the TFD: if reasoning depended so heavily on context, in what sense could one talk about 'general reasoning skills'?

In summary, psychological research for much of the twentieth century has been extremely unsympathetic to the TFD. The majority view, based on Thorndike's original work and its replications, and later research within the cognitive paradigm, was that reasoning skills are highly domain specific: if something is learnt in context X, it is very difficult to apply it to context Y.

But in the last few decades this consensus has been challenged. In the late 1980s a group of psychologists, led by Richard Nisbett, reexamined the issue and claimed that some kinds of general reasoning skills, which they called 'pragmatic reasoning schemas' could be developed through teaching. One of the areas Nisbett concentrated on was statistical reasoning. He gave different groups of students problems such as:

> Catherine is a manufacturer's representative. She likes her job, which takes her to cities all over the country. Something of a gourmet, she eats at restaurants that are recommended to her. When she has a particularly excellent meal at a restaurant, she usually goes for a return visit. But she is frequently disappointed. Subsequent meals are rarely as good as the first meal. Why do you suppose this is?[60]

Nisbett found that participants untrained in statistics would tend to respond by concocting causal hypotheses concerning the restaurant industry's high turnover of chefs, or Catherine's unrealistic expectations. In contrast, Nisbett found that those who had received a basic undergraduate education in introductory statistics were more likely to deploy ideas of chance. They might, for example, suggest that perhaps Catherine had just been lucky the first time she had visited. Those who had received graduate level education in statistics, PhD scientists for instance, would typically give high-quality statistical answers, and appeal to the phenomenon of regression to the mean.[61]

---

[60] Nisbett, 2009.
[61] Regression to the mean refers to the fact that if a variable is extreme on its first measurement, it will tend to be closer to the overall average on its second measurement (and vice versa).

One statistical idea Nisbett focused on was the *Law of Large Numbers*, the idea that as the number of trials in an experiment increases, the average result gets closer to the expected value. The law implies, for example, that the proportion of coin tosses that land heads should be closer to one half when 100 coin tosses take place compared to when there are only 10.

In a series of studies, Nisbett and colleagues investigated whether students who received tuition in the law of large numbers were able to apply it to a variety of different contexts. For instance, one of the problems concerned asking students to explain the fact that after the first two weeks of a baseball season the top batter normally has an average of around 0.450, yet no batter has ever completed a season with an average that high. Typical responses to such a question are causal: for instance, a student might hypothesise that after a run of success in the first few weeks pitchers may study the batter's technique and adjust their tactics. But students who had received training on the law of large numbers were more likely to give a statistical answer, by asserting that two weeks was too short a time to produce an accurate average. As well as increasing the likelihood of giving a statistical response, training also increased the quality of the responses.[62] Critically, the training took place in a domain far removed from that of the test items. Nisbett concluded:

> Can you teach people statistical principles that can affect their understanding of everyday life events without having them take hundreds of hours of courses? Yes.[63]

But although he found some successful transfer in the domain of statistics, Nisbett and his colleagues' investigations into logic were much less promising for the TFD. Does receiving training in abstract logic improve performance on the Wason selection task?[64] Perhaps surprisingly, the answer seems to be that it does not make much of a difference.[65] In fact, neither taking an entire undergraduate course in formal logic,[66] nor two years of graduate education in philosophy,[67] seems to improve students'

---

[62]Fong, Krantz, & Nisbett, 1986; Nisbett, Fong, Lehman, & Cheng, 1987.
[63]Nisbett, 2009, pp. 30–31.
[64]Given the disputes, discussed above and in the next chapter, about which response to the selection task should be considered 'correct', it is necessary to clarify that "improve" here means an increase in the number of participants responding with the normatively correct answer.
[65]Cheng, Holyoak, Nisbett, & Oliver, 1986.
[66]Lehman & Nisbett, 1990.
[67]Morris & Nisbett, 1992.

ability to reason about logic problems (or, at least, about Wason selection tasks) phrased in everyday contexts.

Nisbett explained this difference between his findings for statistics and logic by suggesting that his participants already had "rudimentary, intuitive versions of probabilistic and methodological rules, and when we teach them we are improving on rule systems about which they already have some inkling."[68] In contrast, formal logic is highly non-intuitive and therefore much more resistant to development through education.

Overall, Nisbett's findings in the 1980s and 1990s suggest that the psychologists who, earlier in the century, had rejected the TFD may have been overly pessimistic. He suggested that *some* types of general reasoning skills did exist and could be taught. However, when considering the skill most readily associated with the TFD in the context of mathematics — logical reasoning — the same pessimistic picture emerged. Like Thorndike before him, Nisbett found no evidence that logical thinking skills could be improved through education or training. However, none of Nisbett's studies focused directly on mathematics tuition, which does seem to leave open the possibility that studying mathematics might develop logical reasoning skills in a manner that studying logic directly does not.

1.3 Summary and Plan

In this chapter, we have introduced the TFD, and discussed how it is viewed by two very different communities: mathematicians and psychologists. The theory, which has been supported by thinkers as varied as Plato, Locke and Vorderman, suggests that studying mathematics is valuable because it helps students develop skills of logical reasoning and problem solving. Despite the overwhelming support of mathematics graduates for the theory, and its regular appearance in policy reports, the bulk of empirical work from educational psychologists indicates we should remain sceptical.

This apparently unnoticed divide between the psychological and mathematical communities on this issue is a big puzzle. None of our interviewees, all influential stakeholders in the UK mathematics community, seemed to be aware of this body of research when asked for their views on the TFD and mathematics. And none of the policy reports discussed earlier caveated their support of the TFD by mentioning any empirical research findings on transfer. So we have a strange situation: large numbers

---

[68]Nisbett, 2009, p. 32.

of mathematics graduates and mathematicians are apparently convinced that studying mathematics develops the general thinking skills of logical reasoning and problem solving, but throughout most of the twentieth century psychologists were apparently convinced that this could not be the case.

Our goal in the remainder of the book is to try to resolve this conundrum. We present a series of studies which will give us some more insight into the status of the TFD as it pertains to mathematics. Our first step, taken in the next chapter, is to consider what kinds of research are needed to examine the TFD, and how students' reasoning behaviour can be appropriately measured.

Chapter 2

Investigating the Theory of Formal Discipline

In the first chapter, we discussed some of the background to the TFD, arguing that although the theory is widely accepted by mathematicians and policymakers, most psychologists regard it with scepticism. Our goal in this chapter is to discuss how the veracity of the TFD could be empirically investigated. This discussion requires us to focus on two main topics. First, we discuss how reasoning can be assessed. If we are to see whether the study of mathematics changes a person's reasoning behaviour, we need a way of measuring that behaviour. Second, we discuss how research studies that address the TFD can be designed. Unfortunately the best research design available — a true experiment where students are randomly allocated to study either mathematics or non-mathematics — would be frowned upon by modern ethics committees, so we are forced to make do with designs that have threats to their validity. In the latter part of this chapter, we discuss these threats and what steps can be taken to reduce them.

2.1 Measuring Reasoning

The TFD suggests that studying mathematics develops general reasoning skills, and especially skills of logical reasoning and problem solving. But what exactly are these skills? And how can we assess them? Over the last thirty or so years a large literature on the psychology of reasoning has been developed which has sought to investigate these questions. Different accounts of human reasoning performance have been proposed, and a wide variety of different tasks have been developed to assess different components of reasoning.[1] If our goal is to empirically test the TFD, we need to select some appropriate tasks from this wide array of possibilities.

---

[1] For a review, see Manktelow, 2012.

Rather than make an arbitrary selection, we decided to ask proponents of the TFD for their views. Upon which specific reasoning tasks, do proponents of the TFD believe performance would be facilitated following advanced mathematical study? We did this as part of the interview study discussed in Chapter 1. Recall that we spoke to eight influential figures from the mathematics education community (listed in Table 1.1). After our general discussion about the TFD, during which all participants strongly endorsed the theory, we asked each participant to look at thirteen different reasoning tasks, together with brief explanations of what the tasks were intended to measure. They were asked to look at each task carefully and state the extent to which they agreed with the statement "this task captures some of the skills that studying advanced mathematics develops". We forced our stakeholders to respond on a five point scale (where 1 represented 'strongly disagree' and 5 represented 'strongly agree').

The thirteen tasks, together with associated explanations, are given in full in Appendix A, and are listed in Table 2.1. As the table shows, there was near unanimous agreement about three of the tasks: the belief bias syllogism task, the conditional inference task, and the Wason selection task

Table 2.1   The tasks shown to participants in the interview study, together with their median responses on the five point scale (higher numbers indicate more agreement that the task is likely to be facilitated by mathematical study). Starred tasks were taken from the Watson–Glaser critical thinking appraisal. The thirteenth task, an item from Ravens' Advanced Progressive Matrices, is omitted from the Appendix for copyright reasons.

Task	Figure	Median
Argument Evaluation Task	A.1	4
Belief Bias Syllogism Task	A.2	5
Cognitive Reflection Task	A.3	4
Conditional Inference Task	A.4	5
Evaluation of Arguments*	A.5	3.5
Interpretation*	A.6	4
Recognition of Assumptions*	A.7	4
Estimation Task	A.8	4.5
Insight Problem Solving	A.9	2
Statistical Reasoning Task	A.10	4
Wason THOG Task	A.11	4
Wason Selection Task	A.12	5
Ravens' Matrices		4

all had medians of 5, the highest possible. This meant that the majority of our participants agreed, to the greatest extent possible, that studying mathematics would facilitate performance on these tasks. We briefly discuss each task in turn.

2.1.1 *Syllogisms and belief bias*

Syllogisms are perhaps most closely associated with the ancient Greek philosopher Aristotle. They are logical arguments where a conclusion is deduced from two premises; a classical example concerns the mortality of Socrates:

> All men are mortal.
> Socrates is a man.
> Therefore, Socrates is mortal.

This is a deductive argument: the conclusion of the syllogism necessarily follows from the premises. It is this characteristic, that no extra information has been added to the premises by stating the conclusion, that distinguishes deductive and inductive arguments. The latter, in contrast, involve going beyond the information in the premises, and therefore rely on probabilistic judgement.

Aristotlean syllogisms involve premises of the form 'all a are b', 'some a are b', 'some a are not b', or 'no a are b'. Thus it is possible to create a large number of syllogisms by varying the premises and conclusions. In fact 256 logically distinct syllogisms can be formed, of which only 24 are valid.[2] Clearly, one way of assessing logical reasoning is to present people with a syllogism and ask them to determine whether or not it is valid. The task can be made more interesting by presenting the syllogisms in a variety of real world contexts.

In the early eighties, a team of researchers led by Jonathan Evans, an influential psychologist from Plymouth University, investigated how people's perceptions of the validity of syllogisms interacted with their belief in the accuracy of the syllogisms' conclusions.[3] For instance, consider the

---

[2]These figures depend somewhat on how the syllogisms are counted, and what validity means. For instance, some people suggest that so-called weak conclusions (deducing, for example, that 'some a are b' from 'all a are b') are logically valid, but not psychologically valid.

[3]Evans, Barston, & Pollard, 1983.

following two syllogisms used in Evans's study:

> No millionaires are hard workers.
> Some rich people are hard workers.
> Therefore, some millionaires are not rich people.

> No addictive things are inexpensive.
> Some cigarettes are inexpensive.
> Therefore, some addictive things are not cigarettes.

Most people quickly decide that the first is invalid. But the seconds tends to induce more debate. In fact these two arguments have identical structure (replace 'hard workers' with 'inexpensive', 'rich people' with 'cigarettes', and so on). The only difference is that the first syllogism has an unbelievable conclusion, whereas the second has a believable conclusion. Evans and his colleagues found that participants endorsed the validity of only a third of syllogisms with unbelievable conclusions, compared to 80% of those with believable conclusions. Interestingly, the effect of believability tends to be much stronger on invalid syllogisms than on valid syllogisms. The percentage of people accepting each type of syllogism in the study is shown in Table 2.2. The effect has been replicated many times.[4]

Various accounts of these findings have been proposed.[5] Perhaps the most popular is based on what has become known as dual process theory, an idea about reasoning that we will revisit throughout the book. The theory suggests that there are two distinct types of cognitive process that underlie human reasoning. The two types have various names: some authors refer to them as intuitive and analytic processes, some authors talk about System 1 and System 2, and others prefer to talk about Types rather than

Table 2.2   The belief bias effect found by Evans, Barston and Pollard (1983). Figures give the percentage of participants endorsing the syllogisms as valid.

	Valid	Invalid	Overall
Believable	89	71	80
Unbelievable	56	10	33
Overall	73	41	

---

[4]Morely, Evans, & Handley, 2004; Sá *et al.*, 1999.
[5]For an excellent review, see Manktelow, 2012.

Systems. Here we adopt this latter terminology, and refer to Type 1 and Type 2 processes.[6]

The basic idea is that some cognitive processes are fast, intuitive and take little or no mental effort to perform. These so-called Type 1 processes are extremely useful in day-to-day life. They allow us to, for example, almost instantaneously determine where a sound is coming from, to know which of two objects is further away, to rapidly decide whether or not an object is a threat and, if necessary, to automatically react with an appropriate reflex. Other cognitive actions, known as Type 2 processes, are conscious, slow and effortful. They require attention, and therefore become harder when you are distracted. Examples of Type 2 processes include solving a complicated algebra problem, reading a bus timetable, or planning a holiday.[7]

How does dual process theory help explain belief bias on the syllogism task? The idea is that whenever you see a believable conclusion your Type 1 processing endorses it; it is, after all, consistent with your existing beliefs and knowledge. Similarly, Type 1 processes reject unbelievable responses. This gives you a default response, a kind of gut reaction, which you will go with unless you engage in some effortful Type 2 processing. But, the account suggests, the effect of your Type 1 default response will also influence the nature of your Type 2 processing. Your Type 2 processing is biased towards trying to rationalise your default response. So if your default response is telling you the conclusion is true, you will seek to find a case that satisfies the syllogism. In contrast, if your default response is telling you that the conclusion is false, you will seek to find a case which fails to satisfy the syllogism. Importantly, valid syllogisms are true for all possible cases, whereas invalid syllogisms are typically 'true' for some cases and 'false' for others. So it is much easier to find consistent or inconsistent cases when the syllogism is invalid, and so we would expect the effect of believability to be stronger on invalid syllogisms. As shown in Table 2.2, this is exactly what is normally found.

So the syllogism task can be used to assess reasoning behaviour in at least two different ways. First, one can simply count the number of correct answers a person gives to a variety of syllogisms: how many valid syllogisms they endorse, and how many invalid syllogisms they reject. But second, one can use the task as a measure of the extent to which a person is able to

---

[6]Evans & Stanovich, 2013.

[7]For a popular account of dual process theories of reasoning and decision making, see Kahneman, 2011, or Stanovich, 2004.

reason independently of their prior beliefs. By subtracting the proportion of correct answers they give to syllogisms with unbelievable conclusions from the proportion of correct answers they give to those with believable conclusions, we can construct an index of 'belief bias', the extent to which their prior belief influences reasoning behaviour.

All but one[8] of the respondents in our interview study, strongly agreed with the suggestion that an advanced education in mathematics would assist students in decoupling their reasoning from their prior beliefs. Participant 6, a senior university manager, described the task in these terms:

> What's represented here is a fundamental mathematical skill: being able to start from some axioms and deduce the veracity or otherwise of statements that follow, that are related to those axioms.

So the belief bias syllogism task looks like it would be a plausible task with which to test the TFD. If the TFD were correct (or, at least, if the version of it subscribed to by our interviewees were correct), we would expect students studying advanced mathematics to improve their overall performance on the task, and for the level of their belief bias to decline.

2.1.2   *Conditional inference*

When asked to think of logical reasoning, the statement that springs most naturally to mind is probably the conditional 'if p then q'. This simple sentence has spawned an enormous research literature, and has attracted the attention of psychologists, philosophers and linguists alike. But what does a conditional statement like 'if p then q' mean? Unfortunately, this simple question does not have a simple answer. Instead of offering a definition we start by giving four simple problems from the conditional inference task, a task most associated with Jonathan Evans, whose work on belief bias we have already discussed.

Suppose you are shown a pack of cards, each of which has a letter and a number on. You are then told two pieces of information about a single card, which you must assume to be true, and you are asked to determine whether or not a conclusion follows logically. Must the conclusion necessarily be true in these four cases?

> If the letter is A then the number is 8.
> The letter is A.
> Therefore, the number is 8.

---

[8]One respondent said he was unsure.

If the letter is T then the number is 3.
The number is 3.
Therefore, the letter is T.

If the letter is H then the number is 1.
The letter is not H.
Therefore, the number is not 1.

If the letter is B then the number is 6.
The number is not 6.
Therefore, the letter is not B.

These four problems represent the four conditional inferences that are most often studied by psychologists of reasoning. Almost everyone, around 97%, says that the first is valid. This inference is known as *modus ponens* (MP), and seems straightforward. Slightly over half of us, 56%, typically endorse the second inference, known as *denial of the antecedent* (DA). The third, the *affirmation of the consequent* (AC) inference, is endorsed slightly more often, by around 64% of participants. The final inference, known as the *modus tollens* (MT) inference, is endorsed by around three-quarters of participants, 74%.[9]

There are at least four common ways of interpreting a conditional statement like those in the problems above. The interpretation taught in introduction to logic classes is known as the material conditional. Under this interpretation MP and MT are valid, whereas DA and AC are fallacies. Figure 2.1(a) shows why. Under the material conditional the rule 'if p then

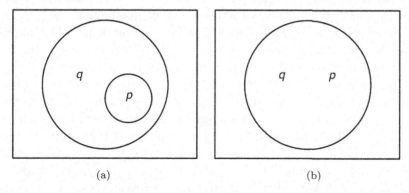

(a)   (b)

Figure 2.1   A diagram illustrating the material conditional 'if p then q' (a) and the biconditional 'p if and only if q' (b).

---

[9]Schroyens, Schaeken, & d'Ydewalle, 2001.

q' means that whenever we have p we must also have q. Formally, 'if p then q' is identical to saying 'q or not-p'. This is illustrated by the p circle, which represents all situations when p is true, being inside the q circle. So, everything inside the rectangle is either an instance of q, or an instance of not-p. This representation suggests why the AC inference is false: knowing q does not tell you p, we could be inside the q circle but outside the p circle. The DA inference is also false: if not-p (outside the p circle), either q or not-q could be the case, all we know is that we are outside the p circle, not whether we are inside or outside the q circle. The diagram also suggests why the MT inference must be true: if not-q, then we are somewhere outside the q circle, which means that we must also be outside the p circle and therefore that not-p must be true.

One of the oddities of the material conditional comes when the statement p is false. Formally in this situation the statement 'if p then q' is true regardless of what q is. For instance, technically 'if France is a monarchy, then $\sqrt{40} = 2$' is a true, albeit unhelpful, statement. One way to think about this is to notice that, because of the MT inference, 'if p then q' is logically equivalent to 'if not-q then not-p', a form known as the contrapositive. Since we know that not-p is true, the identity of q is irrelevant, it could be true or false.

Although the material conditional is typically seen as the normatively correct interpretation of 'if p then q', as noted above, much research has found that most people do not spontaneously adopt it. If they did, they would endorse the MT inference and reject the DA and AC inference, but this pattern of responses is fairly rare.[10] Here we briefly mention three other ways in which the statement 'if p then q' can be interpreted.

Some people interpret 'if p then q' to be a biconditional, which means that they think it means both that p implies q, but also that q implies p. A diagram showing this configuration is given in Figure 2.1(b). Under this interpretation, p and q essentially just mean the same thing. All four inferences discussed above are valid under a biconditional interpretation: if p is true q must be true and vice versa, and if p is false then so must q be, and vice versa. In formal logic, interpreting the statement 'if p then q' as a biconditional is simply a mistake. But it is an understandable mistake as many everyday statements carry this meaning. For instance, imagine a parent telling their child that "if you tidy your room then you can meet your friends tonight". What do they mean? They are trying to establish

---

[10]Evans *et al.*, 1995; Schroyens *et al.*, 2001.

an equivalence between going out and tidying the room. What they really mean is "you can go out if and only if you tidy your room", which is a biconditional.

Another interpretation is what Peter Wason called the defective conditional. This is a slightly unfortunate name, as it seems to suggest that the interpretation is deeply flawed, which is not necessarily the case. Others have adopted more neutral terms, such as the hypothetical conditional,[11] but Wason's term is the more common and is the one we shall adopt here. Under a defective interpretation the sentence 'if p then q' is identical to the material conditional when p is true, but is irrelevant when p is false. In other words unless p is true, no information is added by saying 'if p then q'. This gets round the strange fact, discussed above, that a false p logically implies (via the material conditional) anything. A false p means that the defective conditional is simply irrelevant, it adds no information.

The final interpretation we discuss here is probably the least common. Under the so-called conjunction interpretation, 'if p then q' is taken to mean that p and q always go together. Reasoners who adopt this interpretation believe that the MP and AC inferences are valid (i.e. that p implies q and q implies p) but that the DA and MT inferences are invalid (neither not-p nor not-q allow you to conclude anything).

In the mid-nineties Jonathan Evans and colleagues[12] introduced the negations paradigm, where a "not" is inserted into each component of the conditional 'if p then q'. This gives you four different conditionals: 'if p then q', 'if p then not-q', 'if not-p then q', and 'if not-p then not-q'; and the above discussion can be repeated for each. For instance, drawing the MT inference for the conditional 'if p then not-q' would involve concluding not-p from the premise q. Table 2.3 summarises the identity of the four inferences for each of the four conditionals.

So in the negations paradigm, there are a total of 16 separate inferences that an experimental participant can be asked to endorse or reject. And the situation can be confused further by noting that negative premises can be phrased either explicitly or implicitly. For instance, in the context of letter/number cards, the premise 'not A' could be phrased implicitly as, for example, 'B'. Thus by doubling the number of problems to 32, each negative premise can be phrased both implicitly and explicitly.

---

[11]Mitchell, 1962.
[12]Evans *et al.*, 1995.

Table 2.3   The four inferences (MP, DA, AC and MT) with and without negated premises (Prem) and conclusions (Con). The lower half of the table shows which inferences are valid under which interpretation of the conditional.

	MP		DA		AC		MT	
	Prem	Con	Prem	Con	Prem	Con	Prem	Con
If p then q	p	q	not-p	not-q	q	p	not-q	not-p
If p then not-q	p	not-q	not-p	q	not-q	p	q	not-p
If not-p then q	not-p	q	p	not-q	q	not-p	not-q	p
If not-p then not-q	not-p	not-q	p	q	not-q	not-p	q	p
	MP		DA		AC		MT	
Material	valid		invalid		invalid		valid	
Defective	valid		invalid		invalid		invalid	
Biconditional	valid		valid		valid		valid	
Conjunction	valid		invalid		valid		invalid	

By running experiments with this structure — where participants are asked about implicitly and explicitly negated premises, and with rotated negations in the conditional — two interesting effects have been identified. The first, the negative conclusion effect, is the observation that participants are more likely to endorse conditional inferences with negative conclusions than they are those with positive conclusions. Consider for instance the following two MT inferences:

If the letter is D then the number is not 8.
The number is 8.
Therefore, the letter is not D.

If the letter is not D then the number is 8.
The number is not 8.
Therefore, the letter is D.

The first inference results in a negative conclusion ("the letter is not D") and is more often endorsed compared to the second inference which has a positive conclusion ("the letter is D"). The negative conclusion effect is robustly observed on DA and MT inferences, sometimes on AC inferences, but rarely on the simple MP inference.[13] Although there are at least two different accounts of the negative conclusion effect, we favour the double negation account, which is based on assumptions about Type 2

---

[13] Evans, Handley, Neilens, & Over, 2007.

processing.[14] The account suggests that the reason for the effect is that the sequence of deductive steps required in the first inference above is simply shorter than in the second. Compared to the first inference, the second requires an extra step:

> If the letter is not D then the number is 8.
> The number is not 8.
> Therefore, the letter is not not D.
> Therefore, the letter is D.

According to the double negation account, this extra step, of converting not-not-p into p requires significant Type 2 processing capacity, and it is this necessary extra effort which suppresses endorsement rates.

The second consistent effect found on conditional inference tasks is known as the implicit negation effect, which is sometimes referred to as the affirmative premise effect. This is the finding that participants are more likely to endorse inferences with explicit negations in the premise than those with implicit negations. For instance, the first of the following two inferences is typically endorsed more often than the second, despite both being valid MT inferences:

> If the letter is F then the number is 9.
> The number is not 9.
> Therefore, the letter is not F.

> If the letter is F then the number is 9.
> The number is 7.
> Therefore, the letter is not F.

In contrast to the double negation account of the negative conclusion effect, the implicit negation effect is widely thought to relate to Type 1 processing. According to this account, the term 'not 9' is automatically assumed by Type 1 heuristics to be relevant to the rule 'if F then 9', simply because both contain the symbol '9'. Essentially 'not 9' is a claim about nineness. In contrast, the term '7' is not automatically seen as relevant to the rule. It is a claim about sevenness, not about nineness. Consequently, there is a greater chance of the connection between the two premises being missed, and therefore the argument being rejected. This account has much in common with the matching bias phenomenon on the abstract Wason selection task, which we will encounter in the next section.

---

[14]Attridge & Inglis, 2014; Evans *et al.*, 1995; Evans & Handley, 1999.

So far we have primarily talked about conditionals that refer to abstract content: cards with letters and numbers on. However, of course it is possible to ask about conditionals in context: where p and q refer to the real world. When real world content is allowed into conditionals things become extremely complicated extremely quickly. For instance, the psychologists Phillip Johnson-Laird and Ruth Byrne have suggested that there are at least ten different types of contextual conditionals.[15] For instance, if I tell you that "if you're right, then I'm a monkey's uncle", I'm intending to draw attention to the fact that you're wrong, I'm not suggesting that me being a monkey's uncle is conditional on your being right. So care must be taken to avoid including these kinds of non-standard conditionals when designing contextual conditional inference tasks.

Every one of our interviewees believed that studying advanced mathematics would facilitate more normative behaviour — more behaviour consistent with a material conditional interpretation — on the conditional inference task. All but one interviewee gave the task the highest possible rating on the five-point scale. One participant captured the views of the group when he said:

> [Conditional inference] is something that really is very much in line with the sorts of thinking that math is seeking to deliver [. . .] I think that studying math would put you at a significant advantage [on this task].

So, like the belief bias syllogism task, it seems that the conditional inference task will be helpful for our investigation into the TFD. It seems to capture the kinds of reasoning that those who endorse the TFD believe that studying mathematics develops. The TFD proponents would predict that studying mathematics would reduce the frequency of the behaviour consistent with the biconditional, conjunction and defective interpretations of the conditional, and increase the frequency of behaviour consistent with the material interpretation.

2.1.3   *Wason's selection task*

The third task that over half our participants strongly believed would be facilitated by mathematical study was the Wason selection task, which we have already encountered in Chapter 1. A standard abstract version of the task is given in Figure 1.3.

---

[15] Johnson-Laird & Byrne, 2002.

Those who are familiar with the enormous psychological literature on the selection task will be astonished to discover that in 2003 a version of the task was set as an A level mathematics examination question.[16] Reflecting on students' performance on the question, the chief examiner wrote:

> "[This question] is an old chestnut, and it was very surprising just how completely it floored the candidature. Only a very small number of candidates produced the completely correct answer, which is rather surprising".[17]

If the chief examiner had been more familiar with the psychology of reasoning literature he would have been less surprised. The Wason selection task is probably the most investigated reasoning task in the history of psychology, and participants consistently answer it incorrectly.

As noted in Chapter 1, the normatively correct answer (to the version given in Figure 1.3) is to select the D and 7 cards, because there might not be a 3 on the back of the D, and there might be a D on the back of the 7. Consistently less than 10% of the general population make this selection. The most common answer, which is how around 56% of participants respond, is to choose the D and 3 cards; or sometimes the D card alone, an answer chosen by around 22% of participants.[18] Looking at the individual cards, we find that a large majority, 89%, select the D; a majority, 62%, select the 3; and few people select the 7 (25%) or K (16%) cards.[19]

Wason's original explanation for this finding was that participants were trying to confirm the rule rather than test it. But an important finding that challenged this interpretation emerged a few years later. It was found that if the rule is changed from, say, 'if D then 3' to 'if D then not-3', then the proportion of participants selecting the correct answer — now D and 3 — rose substantially.[20] Moreover, the 'standard' mistake of selecting the p and not-q cards (normally D and 3, but D and 7 in this modified version of the task) became extremely rare. If participants were merely trying to confirm the rule in the task, as Wason suggested, they would surely repeat the same mistake. By rotating the presence of negations in the conditional statement it can be shown that participants seem to have a bias towards selecting the cards mentioned in the rule regardless of the logical validity of

---

[16]Recall that the A level is the qualification taken by 18-year-old school leavers in England and Wales.
[17]OCR, 2003.
[18]Klauer, Stahl, & Erdfelder, 2007.
[19]Oaksford & Chater, 1994.
[20]Evans & Lynch, 1973.

doing so. Both the conditionals 'if D then 3' and 'if D then not-3' mention the D and 3 cards, and these are the cards which tend to be selected. This matching bias seems to be a similar phenomenon to the implicit negation effect discussed in the context of the conditional inference task. Both the statement '3' and 'not 3' are statements about threeness, so the 3 card appears to be relevant to both 'if D then 3' and 'if D then not-3'.

There are many different accounts which try to explain behaviour on the abstract selection task, not to mention the various contextual versions that have been developed.[21] An extensive review would take us too far away from our primary purpose, so here we briefly mention only the dual process account. The basic idea is that Type 1 processes bias your attention towards the apparently relevant parts of the environment. In the standard version of the selection task, this is the D and 3 cards. This creates a kind of default response which Type 2 processes attempt to rationalise. Since this default response appears fairly reasonable, especially if you adopt a biconditional interpretation of the conditional, most people simply accept it. Some people might realise that the 3 card is not needed and reject it from the default response, but it seems that few ever give the cards not included in their Type 1 default response serious attention. Indeed, this has been tested experimentally by recording participants' eye movements: many people simply do not look at the K or 7 cards, and therefore presumably do not even consider whether or not they should be selected.[22]

According to our interview participants, the Wason selection task is also likely to be facilitated by advanced mathematical study. One participant, for instance, said that

> I would be very disappointed if maths students aren't doing much better, maths students at either A Level or university, aren't doing better on [this task].

So the TFD would predict that advanced mathematical study would increase the frequency with which participants select the normatively correct answer, D7, and reduce matching bias. It, therefore, also seems a reasonable task with which to investigate the TFD.

2.2   Incorrect Norms? And Criticisms of Transfer Research

In the first chapter, we discussed Thorndike's work from the first half of the twentieth century. Thorndike's main contribution was to question the

---

[21] Manktelow, 2012.
[22] Evans & Ball, 2010.

idea of transfer: that learnt skills in one domain can be unproblematically applied in another. Thorndike and his followers' approach to studying transfer was straightforward. They would teach a particular skill in one domain, and then test it in a new domain. Clearly, this new domain must share at least some perceived characteristics with the old domain otherwise the study would not make sense: one would not seek to investigate whether teaching learners to conjugate verbs would improve their ability to catch baseballs. This line of thought leads to an important criticism of traditional transfer research: classical transfer studies privilege the perspective of the experimenter. It is the experimenter who designs the study, and therefore who defines what the 'correct' mapping between the learning and test domains is. Therefore, it is the experimenter who decides whether or not transfer has occurred. The mathematics education researcher Joanne Lobato summed up the problem like this:

> Classical transfer studies privilege the perspective of the 'expert researcher' as the standard by which performance is judged. What constitutes transfer must match the 'right' researcher defined solution.[23]

This can result, according to Jean Lave, in an "unnatural laboratory game in which the task becomes to get the subject to match the experimenter's expectations".[24] This criticism essentially suggests that experimenters are applying an inappropriate normative framework to an experimental setting: if the link between the learning and test contexts is not visible to the participant, then the issue of transfer does not arise.

As a result of this criticism Lobato developed what she called the actor-oriented transfer perspective. Under this view, transfer should not be seen as the application of knowledge or skills learnt in one context to a new context, but rather as the generalisation of learning. Instead of seeking to determine whether or not participants in transfer studies correctly identify the link between the learning and test contexts identified by the experimenter, behaviour in novel contexts is scrutinised for evidence of the influence of prior learning.

Somewhat analogously, some critics have suggested that psychologists have applied incorrect norms to some reasoning tasks.[25] Indeed, some went as far as describing certain reasoning tasks, notably Wason's selection task,

---

[23]Lobato, 2006, p. 434.
[24]Lave, 1988, p. 20.
[25]Lopes, 1991.

as being nothing more than "cognitive illusions".[26] Others went further, and suggested that applying the norms of logic to judge participants' behaviour on these tasks is seriously flawed. One critic wrote that "Perhaps the only people who suffer any illusion in relation to cognitive illusions are cognitive psychologists."[27] In response to this kind of view, the future Nobel prizewinner Daniel Kahneman suggested that such critics have "a handy kit of defences that may be used if [participants are] accused of errors: temporary insanity, a difficult childhood, entrapment, or judicial mistakes — one of them will surely work, and will restore the presumption of rationality."[28] Notwithstanding this heated rhetoric, there are coherent arguments against the standard normative model of the conditional, which if accepted would have implications for what the 'correct' response to some reasoning tasks is. One argument, developed and expanded by the psychologists Mike Oaksford and Nick Chater, can best be explained via a philosophical observation about ravens.

Soon after the second world war, the German philosopher Carl Gustav Hempel wrote an influential paper which asked people to consider what would constitute good evidence for the claim that all ravens are black. The obvious intuitive strategy for a would-be verifier would be to go around the world looking for ravens, and checking to see whether those that they found were black. Clearly, the claim can be phrased as a conditional statement: 'all ravens are black' is equivalent to 'if it is a raven, then it is black'. And, as Hempel pointed out, according to formal logic, the statement 'if p then q' is logically equivalent to the contrapositive statement 'if not-q then not-p'. Therefore, saying "all ravens are black" is logically equivalent to saying that "all non-black things are non-ravens". Presumably, therefore, it would also be sensible to go around looking for non-black things, and checking that they were not ravens. But would the observation of a red bus or a blue shoe really constitute evidence in favour of the hypothesis that all ravens are black?

The solution to this problem, which became known as Hempel's paradox, is to realise that ravens are relatively rare, as are black things. In contrast, non-ravens are extremely common, as are non-black things. By formalising a notion of 'information gain', it transpires that the intuitive feeling, which we all share, that searching for ravens is likely to lead to more

---

[26] Cohen, 1981.
[27] Ayton & Hardman, 1997, p. 45.
[28] Kahneman, 1981, p. 340.

useful information about the claim than searching for non-black things can be mathematically justified.

Considerations such as these led Oaksford and Chater to develop a novel analysis of the selection task.[29] They suggested that participants saw the D card (in the rule 'if D then 3') as being one card out of a possible 26 (the numbers of letters in the alphabet), and the 3 card as being one out of a possible 10 (the digits 0–9). Given certain reasonable assumptions, they demonstrated that turning over the 3 card can be shown to yield more information about the truth of the rule than turning over the 7 card. From the perspective of evolutionary adaptivity it makes some sense to evaluate behaviour not by the rules of logic, but by whether behaviour will maximise an organism's chance of survival. Chances of survival, of course, are greatly increased when more information is available about the environment in which the organism is operating. So, under this normative framework, the 'correct' answer is to select the D and 3 cards, not the D and 7 cards, because it is the D and 3 that maximise information gain. Similarly, the appropriateness of the traditional norms applied to the conditional inference task can be criticised on similar grounds.[30]

However, there are those who defend the appropriateness of traditional norms. Keith Stanovich, an influential educational psychologist from the University of Toronto, provided perhaps the most persuasive arguments. He suggested that models of reasoning behaviour can be one of three types: descriptive, normative or prescriptive. A descriptive model describes how humans actually behave, and is simply a matter of empirically investigating reasoning behaviour. In contrast, normative models state how humans ought to reason. However it could be, given that in any given situation we have limited time and limited intelligence, that it would be unreasonable to expect us to act in a normative way. This gives rise to the idea of a prescriptive model, which specifies the best we can hope to achieve given the constraints that we must work within.

Stanovich suggested that there are three different perspectives on reasoning 'errors', which he illustrated using a diagram similar to Figure 2.2. 'Panglossians', like Oaksford and Chater, do not see that there is any substantial gap between the three models.[31] They argue that humans reason as well as they can, and as well as they should. The apparently

---

[29] Oaksford & Chater, 1994.
[30] Oaksford & Chater, 2007.
[31] Stanovich, 1999.

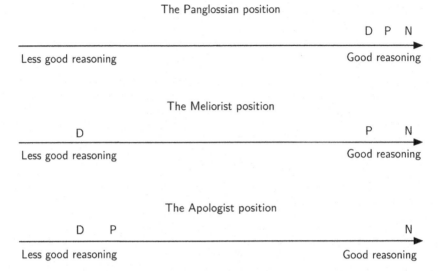

Figure 2.2 The three positions on human rationality identified by Keith Stanovich.
N = normative model, P = prescriptive model, and D = descriptive model.

non-normative responses described above are explained as either random performance errors, incorrect norm application on the experimenter's side, or a misunderstanding of the problem by the participant, due to the experimenter being unclear.[32]

'Meliorists', on the other hand, argue that the way we actually reason is far from the standard by which we could and should reason. They see the prescriptive model as close to the normative model, while the descriptive model falls somewhat short of the other two.[33] The TFD is explicitly a Meliorist position. It states that human reasoning can be improved by mathematical study; in other words that the gap between descriptive and prescriptive models of reasoning can be closed, and will be closed, by studying advanced mathematics.

The final position is that of the 'Apologist'. Apologists agree with Meliorists that the descriptive model falls short of the normative model, but they differ in that they place the prescriptive model much closer to the descriptive. Apologists argue that while we are not reasoning perfectly, we are probably doing the best we could hope to do given our cognitive

[32]Stanovich & West, 2000.
[33]Stanovich, 1999.

limitations.[34] Like the Panglossians, the Apologists believe that we cannot be too critical of human reasoning errors, because we are doing just about as well as we could be. The difference between the two is that the Apologists recognise that this is not up to ideal/normative standards, while the Panglossians argue that it is.

Stanovich argued that there are difficulties with both the Panglossian and Apologist positions. A common Panglossian argument is that the apparent gap between normative and descriptive models is due to random performance errors by the participants — perhaps they are momentarily distracted, for example. The problem that Stanovich identified with this account is that errors seem to be systematic, not random.[35] People consistently make the same mistakes, and the extent of their errors on one task predicts the extent of their errors on other tasks. Performance is also related to cognitive and personality variables, such as intelligence (or cognitive capacity), which seems difficult to account for if errors were merely random.

The incorrect norm argument places the blame for the normative/descriptive gap with the experimenter. Stanovich countered this view by suggesting that we should evaluate reasoning by considering the behaviour of the subset of participants who have the highest levels of intelligence. Intelligence can productively be thought of as a consistent ability for effective behaviour across a wide variety of different environments and situations.[36] This suggests that the behaviour of the most intelligent individuals reflects more effective behaviour than that of less intelligent individuals. When the behaviour of the most intelligent individuals is in line with the normative model set by experts, perhaps this is evidence that the normative model is in fact appropriate. If, on the other hand, the most intelligent individuals usually give a non-normative response, then maybe we should consider revising the normative model as some have suggested.

Stanovich argued that the first scenario is usually the case.[37] He showed that across a variety of tasks, participants were consistent in whether they gave the normative or non-normative response and that there were significant correlations between normative performance and measures of general intelligence. He argued that if experts and more capable participants

---

[34]Stanovich, 1999.
[35]Stanovich, 1999.
[36]Larrick, Nisbett, & Morgan, 1993.
[37]Stanovich, 1999.

agree that the normative response is the correct response, then it very likely *is* the correct response, and this was the case in the majority of tasks he examined. However, there were a small minority of tasks where this was not the case and in such cases, perhaps there is room for the incorrect norm argument to explain the discrepancy between normative and descriptive models.

How serious for TFD research is the actor-oriented critique of transfer, and the related debates about what the appropriate norms are for assessing reasoning performance? In some sense we are able to sidestep the issues because, in contrast to traditional transfer research, the norms are part of the hypothesis that is being tested. Regardless of whether you personally adopt a Panglossian, a Meliorist or an Apologist perspective, the TFD is an explicitly Meliorist position. It predicts that mathematical study will develop students' reasoning behaviour so that it more closely resembles formal logical norms. These norms may or may not be appropriate, but that is a separate issue. For the empirical researcher it suffices to investigate whether or not mathematical study changes reasoning behaviour and, if it does, to assess whether it changes in the direction predicted by the TFD (towards the standard normative models).

Our goal in the next section is to discuss what form investigations in this direction should have. In particular, we discuss issues of research design and ask what research studies are best suited to testing the TFD.

2.3   A Strategy for Testing the Theory of Formal Discipline

In the last section, we discussed various ways of measuring the kinds of reasoning that the TFD proposes that studying advanced mathematics will develop. Our conclusion was that three tasks seem particularly appropriate for this task: the Wason selection task, and tasks concerning syllogistic and conditional reasoning. But of course selecting a reasoning measure is only half the story, we also have to design a programme of research appropriate for testing the TFD. What form should this research take?

Recall that the TFD makes a causal claim: that studying advanced mathematics will cause the development of reasoning skills. Certainly then, we would expect those who had studied advanced mathematics to respond differently to those who had not on each of our reasoning tasks. If that turned out not to be the case it would be highly problematic for the theory. Therefore, a sensible first step in testing the TFD would be to simply determine whether or not there are differences in reasoning behaviour between two groups: those who have, and who have not, studied advanced mathematics. We report investigations along these lines in Chapter 3.

But simply demonstrating a between-groups difference is not sufficient to distinguish between the TFD and its rival, the filtering hypothesis. What if those students who already show particular patterns of reasoning are simply more likely to choose to study advanced mathematics? How can we establish a causal relationship between mathematical study and reasoning development?

There is a standard method of establishing causality in behavioural sciences such as psychology, medicine and education, but unfortunately it cannot be used to test the TFD. So-called true experiments work like this: we have some intervention that we want to show causes some outcome on some population. In our case the intervention would be post-compulsory mathematical study, and the population might be, say, all 16-year-old students. We find a random sample of this population, and randomly allocate each member of the sample to either receive, or not receive, the intervention. The random nature of this allocation is crucial. It guarantees that there are only random differences between our experimental and control groups before the intervention starts. Because we have randomised we know that any preexisting between-groups differences cannot, by definition, be systematic.

Then, after the intervention, we use our outcome measure to see how well each participant is doing. If those who had received the intervention are doing, on average, substantially better than those who did not, we can conclude that this difference is either down to random chance or is down to our intervention. Happily standard statistical methods are able to calculate the probability of our results being down to random chance, and if it is sufficiently low, it seems reasonable to reject the possibility. Assuming our study has been well controlled, this leaves us with only one possibility, that our observed between-groups difference was caused by the intervention.

Unfortunately, the absolutely critical part of this research design is the randomisation into groups. And it is this part that we simply cannot do when investigating the TFD. It would neither be ethical, nor practical, to randomly allocate students' post-compulsory subject choices: which school-leaver would be happy to have their future career prospects constrained by the demands of a TFD researcher?

If we cannot randomise into groups, we cannot conduct a true experiment. We're left with a design known as a quasi-experiment, where the experimental and comparison groups are formed through some process outside of the researcher's control, in our case individuals' post-compulsory study choices. This leaves open the possibility that any observed post-intervention difference is due to pre-existing differences between the groups. Although we cannot rule this possibility out completely, we can take steps to render it less plausible. One obvious method is to conduct a so-called longitudinal study by measuring reasoning development over time, rather than at a single point. We can, for instance, take measures of the reasoning performance of our groups both before and after the intervention. If we find no differences at time 1, then that should count as evidence against the filtering hypothesis. But it is not overwhelming evidence. In Chapter 1, we briefly discussed the so-called Matthew effect, which can be colloquially summarised as 'the rich get richer, the poor get poorer'. Even if our two groups do not differ on our primary reasoning measure at time 1, what if they systematically differ on some other background variable which makes them more likely to develop reasoning skills, independently of their mathematical study?

What kind of background variable might be predictive of reasoning gains, independently of mathematical study? Two obvious contenders arise from Keith Stanovich's analysis of Type 2 thinking processes.[38] Stanovich suggested that there are two major determinants of whether or not someone will be successful at an effortful thinking task. The first is their cognitive capacity: some people are simply more effective at processing complex information than others. These individual differences are indexed well by traditional intelligence tests, which turn out to be predictive of success in very many areas of day-to-day life.[39] However, Stanovich suggested that there is a second important predictor. Although people vary in their ability to process complex information, they also vary in their willingness to do so. This, a person's so-called 'thinking disposition', is also highly predictive of their performance on various reasoning tasks.[40] After all, it doesn't matter if a person is capable of successfully engaging with demanding cognitive tasks if they rarely do so because they don't enjoy it.

How can a person's thinking disposition be measured? Although there are various self-report scales,[41] there are also behavioural tests. The

---

[38] Stanovich, 2009.

[39] Deary, 2001; Flynn, 2009; Judge, Higgins, Thoresen, & Barrick, 1999; Larrick *et al.*, 1993.

[40] Toplak, West, & Stanovich, 2011.

[41] For instance, the 'Actively Open Minded Thinking Scale' asks participants to rate the extent to which they agree with statements such as "Difficulties can usually be overcome by thinking about the problem, rather than through waiting for good fortune".

Cognitive Reflection Test (CRT), which was one of the tasks we asked our interviewees about, has been used for this purpose.[42] The full CRT is shown in Figure A.3, but focusing on the following single question suffices for our purposes here:

> A bat and a ball costs £1.10 in total. The bat costs £1.00 more than the ball. How much does the ball cost? ... pence.

If you are like the majority of people who have been given this question over the years, you will probably be thinking that the answer is ten pence. But in fact this widely shared gut reaction is wrong. The correct answer is five pence: if the ball costs 10 pence, then the bat must cost £1.10, and together they will cost £1.20 not £1.10.

This question, along with two similar problems which together make up the CRT, was first developed by Shane Frederick, a psychologist interested in economic decision making.[43] He was surprised to discover that even very high achieving students at some of the US's best universities struggle with the question: around a fifth of students at Princeton and Harvard failed to get any of the three questions on the CRT correct, and the figure is higher in studies involving samples from the general population.

What makes the CRT so difficult? The answer again relies upon dual process accounts of reasoning: each of the questions on the test suggest a default response which, in each case, is wrong. But only if the test taker is willing to devote effortful Type 2 processing to carefully evaluate their gut reaction will they notice that it is incorrect. Thus the CRT can be used as a measure of a person's willingness to engage in effortful analytic thinking.[44] Such a measure is necessary if we are to investigate whether the TFD might be due to a Matthew effect on thinking dispositions. For example, what if students who choose to study mathematics have systematically different thinking dispositions to those who do not? If so, and if they do show different patterns of reasoning development, maybe this could be due to differences in pre-existing thinking dispositions, not differences in exposure to mathematics education.

To recap, we cannot conduct a true experiment to investigate the TFD for obvious ethical reasons: students' post-compulsory subject choices cannot be allocated randomly. So we are left with quasi-experimental

---

[42] Toplak *et al.*, 2011.
[43] Frederick, 2005.
[44] Toplak *et al.*, 2011.

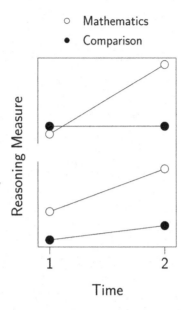

Figure 2.3   Possible outcomes of a two-group longitudinal quasi-experimental design.

designs. As well as testing students' reasoning behaviour before and after the 'intervention' — a course of advanced mathematics — we can take measures of factors which might predict the extent to which their reasoning skills might develop anyway. Stanovich's work suggests two sensible choices: students' cognitive capacity as indexed by classic intelligence tests, and their thinking dispositions, as indexed by self-report measures and the cognitive reflection test. If we found that reasoning changes between Times 1 and 2 were unrelated to either of these factors, that should give us more confidence that any observed between-groups differences we observed was not due to the Matthew effect.

To summarise, two possible results of a two-group longitudinal quasi-experimental study are shown in Figure 2.3. The horizontal axis shows the time at which the measure was taken, and the vertical axis shows the outcome of a given reasoning measure, with better scores higher up the axis.

The result in the top half of the graph is the easiest to interpret. It shows that the mathematics and comparison groups showed roughly similar reasoning behaviour at Time 1, but that by Time 2 they had diverged. This finding, coupled with a lack of association between any background measures and gain scores, would provide fairly strong support for the TFD. But the result shown in the bottom half of Figure 2.3 would be harder to interpret. Although the mathematics group appears to be developing more than the comparison group, the two groups started at different points, and both groups have improved slightly. This pattern of results would be consistent both with the TFD, and with a filtering effect coupled with the Matthew effect. Here it is simply not possible to tell whether the greater gains shown by the mathematics group were due to their mathematical study, or due to their advantage at Time 1 in reasoning skills, or in some background factor associated with reasoning skills.

Nevertheless, although it has weaknesses, given the ethical restrictions of conducting true experiments in this area, a longitudinal quasi-experimental study would be a sensible next step after establishing that there are indeed differences in reasoning behaviours between those who have, and who have not, studied mathematics. We report discuss two such studies in Chapter 4.

2.4 Plan for the Remainder of the Book

Our goal for the remainder of the book is to empirically investigate the TFD using the three tasks discussed in this chapter: the conditional inference task, the syllogistic reasoning task and the Wason selection task. In the next chapter, we directly investigate whether or not those students who have studied mathematics reason differently to those who have not. Then we report two longitudinal quasi-experimental studies designed to investigate whether the pattern of reasoning development followed by mathematics and non-mathematics students is consistent with the TFD. Along the way, we report several other studies which reinforce our findings from the main studies, or which address questions that the main studies left unresolved.

The findings from all these studies paint a more nuanced picture than either the defenders of the TFD or its critics would have predicted. We will argue that studying advanced mathematics does seem to be associated with development in reasoning skills, but that the nature of that development is more limited that the TFD suggests, and that it is not always in the direction predicted by the theory. Nevertheless, we will argue that Thorndike was wrong to dismiss the TFD completely, and that in some important ways, studying mathematics does appear to develop general reasoning skills.

Chapter 3

Cross-Sectional Differences in Reasoning Behaviour

The first step in our TFD-testing strategy is simply to determine whether or not there are group differences in reasoning behaviour. Do people who have studied advanced mathematics perform differently on our three identified tasks — syllogisms, conditional inference and the Wason selection task — to people who have not studied advanced mathematics? If the answer were no, then it would cast substantial doubt upon the TFD.

Here we report two studies, both involving UK-based university mathematics and non-mathematics students. The UK is unusual in that students can, and mostly do, stop studying mathematics at age 16.[1] Those who wish to study mathematically demanding university courses, such as science, engineering or mathematics itself, typically opt at the age of 16 to study a two-year course known as Advanced level mathematics, or A level mathematics. The A level is a demanding programme, which represents between quarter and a third of students' total programme of study. Thus most undergraduate mathematics students in the UK will have studied mathematics for at least two years more than most arts/humanities undergraduate students.

Although there are three different versions of the A level mathematics course available to students in England, all have similar content. The syllabus contains sections on algebra, geometry, calculus, trigonometry, probability, mathematical modelling, kinematics and forces.[2] Most importantly, students are not taught any substantially proof-based mathematics, nor are they taught the definition of the conditional statement. To formally establish this, as well as inspecting the syllabus, we conducted an analysis of every first year A level mathematics examination between 2009 and 2011.

---

[1] Hodgen *et al.*, 2010.
[2] Assessment and Qualifications Alliance, 2015.

Of 929 questions set, only one contained an explicit "if...then" sentence, and there were no mentions at all of the terms "modus ponens", "modus tollens" or "conditional".

3.1   Conditional Inference and Syllogistic Reasoning

In the spring of 2013, we asked a large number of undergraduate students from UK mathematics and English departments to visit a website and answer a series of reasoning problems.[3] The students were contacted via their departmental secretaries and were told that the study was concerned with reasoning behaviour in higher education. They were not informed of our specific interest in the TFD, or that only mathematics and English undergraduates were invited to take part.

Having received an email, willing participants clicked through to the experimental website, gave their consent to participate in the study and provided some basic demographic information: what course they were studying and what year of study they were in.

Participants then were asked to solve several items taken from Raven's Advanced Progressive Matrices. These items are designed to measure non-verbal intelligence. On each item, participants are asked to identify the missing element that correctly completes a pattern. They were given five minutes, to solve up to six items. After five minutes, the website automatically loaded the next page.

As noted in Chapter 2, given the impossibility of conducting a true experiment, attempting to control for existing differences between our two groups is an important part of investigating the TFD. But it is worth noting the weakness of our intelligence measure here: five minutes is a very short test, which we would not expect to provide an accurate measure of intelligence. Nevertheless, given the constraints of a web-based study, and given that this was a preliminary investigation, we felt that this was a reasonable approach. In some of the studies reported later in the book we used more typical measures of non-verbal intelligence.

After participants had spent five minutes on the Raven's items, the next page loaded and the main reasoning task began. At this stage, participants

---

[3]Conducting experimental studies of human behaviour online is now commonplace. Investigations suggest that results from web-based experiments are typically consistent with those from lab-based experiments (e.g. Krantz & Dalal, 2000).

Table 3.1   The four conditional inferences with and without negated premises (Pr) and conclusions (Con).

Conditional	MP		DA		AC		MT	
	Pr	Con	Pr	Con	Pr	Con	Pr	Con
If p then q	p	q	not-p	not-q	q	p	not-q	not-p
If p then not-q	p	not-q	not-p	q	not-q	p	q	not-p
If not-p then q	not-p	q	p	not-q	q	not-p	not-q	p
If not-p then not-q	not-p	not-q	p	q	not-q	not-p	q	p

were randomly split into one of six different groups:

- Those in the *abstract conditional inference* group were given 16 items from Evans's Conditional Inference Task.[4] These were constructed by rotating the presence of negations in the rule 'if p then q', and by asking participants whether MP, DA, AC and MT were valid inferences. Table 3.1 shows how this leads to 16 different problems.[5]

- Those in the *thematic conditional inference* group were given 16 items that were equivalent to those given to the abstract group, except that they contained meaningful content. For example, one question concerned the conditional "if oil prices continue to rise then UK petrol prices will rise" and another the conditional "If car ownership increases then traffic congestion will get worse".[6]

- Those in the *abstract disjunctive inference* group were given the same problems as those in the abstract conditional inference group, but where the conditional statement was replaced with the equivalent disjunctive statement. For example, rather than the problem concerning the conditional statement "if the letter is A, then the number is 3", it concerned the disjunctive statement "the letter is not A or the number is 3". As noted earlier, according to classical logic these statements are logically equivalent. Table 3.2 gives the 16 different problems.

- Those in the *thematic disjunctive inference* group were given the same problems as the thematic conditional group, except with the statements rephrased as disjunctives. For instance, they were asked about the statements "Oil prices will not continue to rise or UK petrol prices will

---

[4]Evans *et al.*, 1996.

[5]The full list of items used is given in Appendix B, they were questions 1–16 from the full version of the task, all of which had explicit negations.

[6]These items were taken from those used by Evans *et al.* (2010). The full list is given in Appendix C.

Table 3.2   The four disjunctive inferences with and without Pr and Con.

Disjunctive	MP		DA		AC		MT	
	Pr	Con	Pr	Con	Pr	Con	Pr	Con
Not-p or q	p	q	not-p	not-q	q	p	not-q	not-p
Not-p or not-q	p	not-q	not-p	q	not-q	p	q	not-p
p or q	not-p	q	p	not-q	q	not-p	not-q	p
p or not-q	not-p	not-q	p	q	not-q	not-p	q	p

rise" and "Car ownership will not increase or traffic congestion will get worse".

- Those in the *abstract syllogism* group were given eight Aristotelian syllogisms with abstract content. For example, in one problem participants were asked to determine whether or not the statement "Alls Ds are Ss" logically followed from the premises "All Ss are Es" and "All Ds are Es".[7]

- Those in the *thematic syllogism* group were given eight Aristotelian syllogisms with identical logical structure to those in the abstract group, but where the content was given in context. To avoid issues related to belief bias, we used content related to alien objects.[8] For instance, one problem asked participants to decide whether one can conclude "Podips are lapitars" from the premises "All lapitars wear clothes" and "Podips wear clothes".

Participants in all groups had as long as they liked to complete their problems, and once they had finished they were thanked for their time and the experiment ended.

A total of 438 students completed the study, 167 in the mathematics group and 271 in the comparison group. We first investigated whether or not the two groups differed on our short intelligence measure. On average the mathematics group answered 3.8 items correctly (out of 6), whereas the comparison group averaged 2.8 correct answers. This difference was highly significant, meaning that is was unlikely to have come about by chance.[9] Given this difference, we controlled for intelligence in our main analyses.

---

[7] These items were adapted from those used by Sá *et al.* (1999), and the full list is given in Appendix D.

[8] The items were taken from those used by Sá *et al.* (1999), the full list is given in Appendix E.

[9] Comparing these means using a Welch's t-test revealed a highly significant effect, $t(268.7) = 8.054, p < 0.001$.

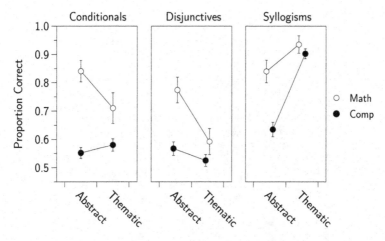

Figure 3.1 The proportion of correct answers given by the mathematics and comparison groups for each of the six sets of problems. Error bars show ±1 standard error of the mean.

Figure 3.1 shows the average proportion of correct responses given by the two groups for each of the six sets of problems. All three sets of problems show a similar pattern: an advantage for the mathematics group on abstract problems, but a reduced advantage for the thematic problems.

A statistical analysis supports this reading of the results, including when differences in intelligence are controlled for.[10] In other words, the mathematics group performed significantly better than the comparison group on all three abstract sets of problems, even when their advantage on the non-verbal intelligence measure had been taken into account.

---

[10]In view of their logical equivalence, the conditional and disjunctive tasks were analysed together using a context (abstract/thematic) by statement-type (conditional/disjunctive) by group (mathematics/comparison) between-subjects analysis of covariance (ANCOVA), where intelligence scores were a covariate. This revealed significant main effects of context ($F(1, 256) = 13.3, p < 0.001$), statement-type ($F(1, 256) = 6.11, p = 0.014$) and group ($F(1, 256) = 42.4, p < 0.001$), and, crucially, a highly significant context by group interaction $F(1, 256) = 9.94, p = 0.002$). Bonferroni-corrected post-hoc t-tests revealed significant group differences for the abstract problems, $ps < 0.001$, for the thematic conditionals, $p = 0.037$, but not for the thematic disjunctives, $p > 0.2$. The syllogism data were analysed with a context (abstract/thematic) by group (mathematics/comparison) ANCOVA with intelligence scores as the covariate. This revealed significant main effects of context ($F(1, 168) = 65.7, p < 0.001$) and group ($F(1, 168) = 17.8, p = 0.001$) and, crucially, a significant group by context interaction ($F(1, 168) = 15.0, p = 0.003$). A Bonferroni-corrected post-hoc test found that the two groups differed significantly on the abstract problems ($p < 0.001$) but not on the thematic problems ($p > 0.8$).

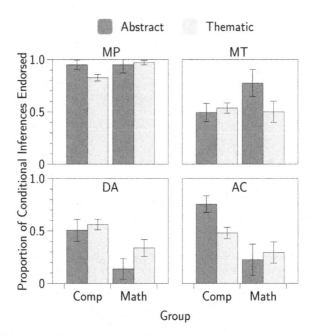

Figure 3.2   The proportion of conditional inferences endorsed by each group. Higher scores are better for MP and MT, lower scores are better for DA and AC. Error bars show ±1 standard error of the mean.

While the mathematics group also performed significantly better than the comparison group on the thematic conditionals, this effect was smaller, and the difference was not significant for the thematic disjunctive or thematic syllogism problems.

Where did the mathematics students' advantage on the abstract problems come from? For the conditional inference problems, we can meaningfully break down the results further. Recall that we asked our participants to evaluate four different types of inference: modus ponens (MP), denial of the antecedent (DA), affirmation of the consequent (AC) and modus tollens (MT). In Figure 3.2, we show the proportion of the different inferences endorsed by those participants who were given tasks involving conditional 'if p then q' statements. Because DA and AC are invalid inferences, a low proportion of endorsements indicates more normatively correct answers in these graphs, whereas for the valid MP and MT inferences, the reverse is the case.

Unsurprisingly, both groups endorsed the straightforward MP inferences nearly all the time. There were group differences on both the invalid inferences, with the mathematics students more likely to reject the DA and AC inferences compared to the comparison group.[11] The MT inference showed a slightly different pattern. On the abstract problems the mathematics group correctly endorsed more inferences than the comparison group, but this was not the case for the thematic problems.[12]

One way of quantifying the size of group differences is to give an 'effect size', which represents the difference between the two groups' scores as a multiple of the overall standard deviation, a measure of the variation in the scores. This, the so-called Cohen's *d*, revealed that the size of the group differences was bigger for the abstract AC and DA inferences, −1.12 and −1.90,[13] compared to the abstract MT inference, +0.96; and that all the abstract effect sizes were larger than all the thematic effect sizes (the equivalent *d*s for the thematic DA, AC and MT inferences were −0.63, −0.48, and +0.10, respectively). In other words, the mathematics group's advantage on these abstract conditional problems was around one standard deviation for the valid abstract MT inference, and even larger for the invalid abstract DA and AC inferences. But this advantage was considerably smaller on the thematic problems, and abolished entirely on the thematic MT problems.

Another way of expressing these data emerges by asking what they tell us about the interpretation of the conditional adopted by the mathematics and comparison groups. In Figure 3.3, we have plotted the extent to which the two groups behaved in accordance with the different interpretations of the conditional introduced earlier in the book. Overall, the mathematics group's responses were more material, more defective and less biconditional than the comparison group's.

When discussing the abstract conditional inference task earlier, we also noted that it could be used to assess the extent to which participants exhibit two distinct reasoning effects: the implicit negation effect and the negative conclusion effect. Because we used a shortened version of the task for our web study (long online studies tend to suffer higher rates of participant drop out) we were unable to calculate measures of the implicit negation effect. However, we were able to assess the extent to which the mathematics and English undergraduates exhibited the negative conclusion

---

[11]Welch's *t*-tests: DA: $t(79.2) = 5.00, p < 0.001$; AC: $t(64.16) = 4.92, p < 0.001$; MT: $t(57.50) = -1.814, p = 0.075$.

[12]There was a significant group by context interaction, $F(1, 124) = 5.71, p = 0.018$.

[13]The minus signs here indicate that the mathematics group had *lower* endorsement rates (i.e. more normative responses) than the comparison group.

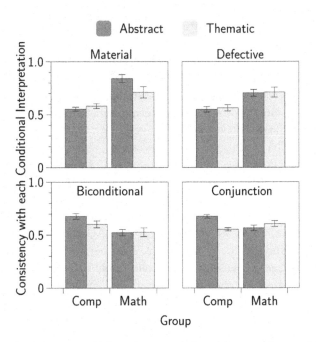

Figure 3.3   The extent to which each group behaved in accordance with the four interpretations of the conditional introduced earlier in the book. Error bars show ±1 standard error of the mean.

effect. This effect refers to the phenomenon that reasoners typically, are more willing to endorse inferences that have negative conclusions than those with affirmative conclusions. This can be assessed simply by summing the number of negative conclusions a participant endorses and subtracting the number of affirmative conclusions they endorse. We calculated this 'negative conclusion index' for each participant who tackled conditional problems. The mean for the English group was +0.54 compared to +0.05 for the mathematics group. Although these indices suggest that the mathematics group appeared to be less biased than the English group — i.e. they endorsed roughly the same number of negative and affirmative conclusions, whereas the English group endorsed slightly more negative conclusions — the difference was small, and not statistically significant[14] meaning that it may not be a robust effect.

What does this set of results tell us about the TFD? Two main lessons can be drawn. First, it clearly was the case that the students in our

---

[14]Welch's *t*-test, $t(76.4) = 1.42, p = 0.158$.

mathematics group reasoned differently to the students in our English group. Largely, their performance was closer to the norms of classical logic, as the TFD would predict. However, the mathematics group's performance was some distance away from ceiling levels: deductive reasoning is hard, and while these findings leave open the possibility that studying mathematics improves students' reasoning, it clearly is not the case that it leads to completely normative behaviour.

Second, this headline finding should be caveated by emphasising that the difference appeared to be substantially larger on reasoning problems that involved abstract content, compared to those that involved thematic content. To illustrate this further, consider the following two MT inferences:

If the letter is D then the number is 4.
The number is not 4.
Therefore, the letter is not D.

If car ownership increases then traffic congestion will get worse.
Traffic congestion does not get worse.
Therefore, car ownership does not increase.

Substantially more students studying mathematics than English endorsed inferences similar to the first, but there was no significant group difference on inferences like the second. This suggests that further investigations about whether or not the mathematics advantage we found in this study is causally related to mathematical study would be most productively focused on problems with abstract content. We explore the critical issue of causality further in the next chapter, but first we consider whether or not there are cross-sectional group differences in behaviour on the third reasoning task identified in Chapter 2, the Wason selection task.

3.2   The Wason Selection Task

As noted in Chapter 2, the Wason selection task has appeared as an A level mathematics examination question. This is really quite peculiar because the task is notoriously difficult: typically, only around 10% of people select the normatively correct answer. Many theories have been proposed to account for this, but they need not concern us here.[15] Instead our focus is simply this: do those who have studied advanced mathematics respond to the Wason selection task differently to those who have not?

---

[15]Interested parties should consult Manktelow's (2012) excellent review.

In collaboration with Adrian Simpson and Derrick Watson, we recruited 457 undergraduates from the mathematics and history departments of a highly ranked UK university. As with the first study reported in this chapter, the study was conducted online. Emails were sent to students from both departments inviting them to take part in the study. Those who agreed clicked through to a website with the following version of the selection task:

> Four cards are placed on a table in front of you. Each card has a letter on one side and a number on the other. You can see:
>
> $\boxed{\text{D}}$ $\boxed{\text{K}}$ $\boxed{3}$ $\boxed{7}$
>
> Here is a conjectured rule about the cards: *"if a card has a D on one side, then it has a 3 on the other".*
>
> Your task is to select all those cards, but only those cards, which you would need to turn over in order to find out whether the rule is true or false.

After submitting their answers participants were asked whether or not they had seen the selection task before, and went on to complete an unrelated task. The few participants who had seen the task before were deleted from the data.

The range of answers given by each group are shown in Table 3.3. The range of responses from the history students was roughly in line with what one might expect from prior research. But a substantially greater number of mathematics students selected the normative answer (18%) compared to the history students (6%).[16] Interestingly, the *range* of answers from the

Table 3.3 The distribution of selections from each group.

	Maths		History	
	Raw	%	Raw	%
D	133	46	64	38
D3	47	16	53	31
D7*	51	18	10	6
DK7	17	6	0	0
D37	10	4	4	2
DK37	20	7	22	13
Other	11	4	15	9
N	289		168	

Note: *Normative answer.

---

[16] $\chi^2(1) = 12.6, p < 0.001.$

Table 3.4 The % of each group selecting each card.

	Maths (u/g)	History (u/g)
D	98	95
K	16	20
3	29	52
7	35	24

mathematics undergraduates was different to those of the historians.[17] As well as being more successful, the mathematics undergraduates seemed to make different mistakes. Of particular interest is that substantially fewer of the mathematics undergraduates selected the matching D and 3 cards (16%), than did the history undergraduates (31%).

Table 3.4 shows the percentage of participants from each group who selected each card, collapsed across selections. There were significant between-groups differences for the 3 and 7 cards, with fewer participants from the mathematical group picking 3 and more picking 7. Although the mathematical sample were generally more successful than the comparison group (and results from the literature), as with the conditional inference and syllogisms tasks reported above, they were not overwhelmingly successful.

Why did relatively few mathematicians make the 'standard mistake' of selecting the 3 card? And why did relatively few successfully pick the 7 card? Recall that one hypothesis that attempts to explain behaviour on the selection task relies on dual process theory. The idea is that Type 1 processes bias attention towards apparently relevant parts of the environment. Some people's attention is directed solely towards the D card because this is the antecedent of the rule 'if D then 3'. Jonathan Evans has referred to this as the 'if heuristic'. Other people's attention is directed towards both the D and 3 cards, because these are the symbols in the rule, a phenomenon known as the 'matching heuristic'.[18] Thus the D card, or the D and 3 cards, form a default response which is rationalised by Type 2 processes unless an error is found. Crucially, because attention is focused on this default response, little cognitive effort is devoted to the K or 7 cards. Indeed, when participants' eye movements are recorded, it transpires that some people barely look at the K and 7 cards at all.[19]

---

[17]$\chi^2(6) = 43.3, p < 0.001$.
[18]Evans, 2006.
[19]Evans & Ball, 2010.

This theory suggests one hypothesis that accounts for the mathematicians' reduced tendency to select the 3 card, but also their relatively low levels of 7 selections. Perhaps, like the general population, mathematicians' attention is biased towards considering only the D card, or only the D and 3 cards, by Type 1 processes, but in addition perhaps they tend to be more successful at determining, with slow Type 2 processes, that the 3 card is not relevant. If the mathematicians typically do not look at the K or 7 cards, this would also account for the relatively low number of D7 responses: the mathematicians in our sample would be responding normatively to the subset of the task that they focused on.

To test this account we ran a second study, again in collaboration with Adrian Simpson and Derrick Watson. This time, 58 undergraduate and postgraduate students from a highly ranked UK university were recruited, 30 of whom were mathematicians and 28 were from the arts faculty. None of the participants had seen the task before, and were paid volunteers, recruited on the basis that they did not wear eye-glasses. The participants tackled the selection task on a computer screen while we recorded their eye movements.[20]

The experiment involved a standard version of the selection task. The display used for the cards is shown in Figure 3.4, although the location of each card was rotated between participants (i.e. for some participants the top left card was a 7, etc.). Participants responded to the task by pressing a button and then announcing their answer out loud. They were encouraged to take as long as they needed.

We first investigated the range of responses given by participants. These were similar to those from the large web-based study: the mathematics group made more 7 selections than the arts group (43% vs. 14%) and fewer 3 selections (36% vs. 43%), although this latter difference was of a lower magnitude than that found in the web study. There was no significant difference between the overall response times of the two groups.

To test our hypothesis — that the mathematics group rejected the 3 card with slow Type 2 processes, whereas they rejected the K and 7 cards with rapid Type 1 processes — we calculated how long each participant spent looking at the cards they ended up not selecting. These figures are

---

[20]We used an SR Research EyeLink I system. This is a head mounted infrared-based eye tracker that automatically compensates for head movements and achieves an average gaze accuracy of approximately $0.5-1.0°$. We did not restrict participants' head movements.

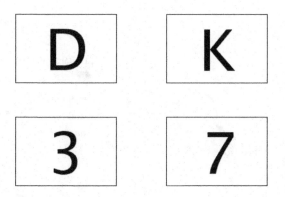

Figure 3.4   The display as seen by the participants.

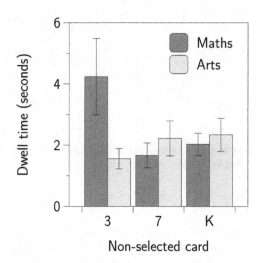

Figure 3.5   The mean dwell times (ms) for the non-selected 3, 7, and K cards. Error bars represent ±1 standard error around the mean.

shown in Figure 3.5 (we excluded the D card from this analysis as almost all participants selected it).

In line with our prediction, we found that the mathematicians who did not select the 3 card spent substantially longer doing so than the arts

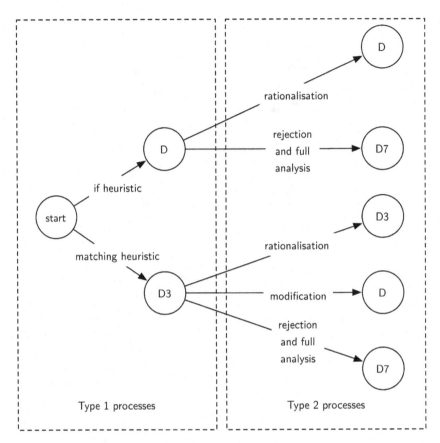

Figure 3.6   The main range of possibilities for a participant when encountering the selection task, according to the dual process account.

students.[21] In contrast, the two groups spent roughly similar times rejecting the K and 7 cards.

Figure 3.6 illustrates what we suspect is going on here. Overall, our data are consistent with the suggestion that an individual's attention is directed towards either the D card, or both the D and 3 cards. In the case of the if heuristic, a default response of D is generated and this is either rationalised and selected, or (rarely) a full analysis is conducted which leads to the normative response. In the case of the matching heuristic, the D and 3 cards are cued. If, during the rationalisation process, the participant

---

[21] A group by card ANOVA was conducted, which revealed a significant group by card interaction, $F(2, 116) = 3.46, p = 0.035$.

notices that the 3 card is not relevant to the rule, they might modify their default response and select just the D card. More mathematicians than arts students appeared to fall into this group, thus explaining their longer mean dwell times on rejected 3 cards.

In other words, both groups rejected the K and 7 cards — those which were not cued as relevant by being in the 'if D then 3' rule — because of rapid Type 1 processes. They simply were not included in participants' default responses, and so were not closely studied using slow Type 2 processes. But there was a group difference on the time spent to reject the 3 card. Our interpretation of this is that many of the mathematicians whose attention was directed towards the D and 3 cards were successful at determining that turning over the 3 card would not assist them with testing the rule.

Overall, our findings with the Wason selection task are consistent with those from the conditional inference task and syllogistic reasoning task reported earlier in the chapter. We found group differences on all three tasks. In each case, those who had studied advanced level mathematics (at least the equivalent of two years of post-compulsory study) gave more correct answers, as defined by norms of classical logic. This is exactly what would be predicted by the TFD.

Two further points are worth highlighting, which will influence our choice of tasks in the studies reported in the next few chapters. First, the mathematics students' performance on the Wason selection task was rather poor. Only 18% of mathematics students selected the normatively correct answer. While this was substantially higher than the 6% in our comparison group, it nevertheless seems inconsistent with a strong form of the TFD. Our eye movement study suggests a reason for this: we found evidence that many in the mathematics group simply did not consider the K or 7 cards at all. Their attention was biased by Type 1 heuristics towards considering only the D and 3 cards. Seen in this light, the relatively low proportion of mathematics students who selected the 3 card (29% compared to 52% of the arts students) seems more in line with what the TFD would predict. The complexity of interpreting behaviour on the Wason selection task might suggest that it is not the best way of investigating the TFD.

Second, we found that the biggest advantage for mathematics students was on abstract tasks, not thematic tasks. While the mathematics students outperformed control groups on the thematic tasks, this advantage was only statistically significant for the conditional problems. In general, it seems that if the TFD is valid its largest impact is upon abstract reasoning skills.

Given this it might be prudent to focus testing the causal aspects of the TFD with abstract tasks alone.

In the next chapter, we focus on the critical developmental hypothesis that underlies the TFD. Specifically, we will try to distinguish between the filtering hypothesis and the TFD. Is it that studying advanced mathematics develops reasoning skills? Or is it that those who already have certain reasoning skills are simply more likely to choose to study mathematics?

Chapter 4

Longitudinal Development in Conditional Reasoning

In Chapter 3, we presented some evidence that groups of students who have studied advanced mathematics reason differently to groups of students who have not. We used the three reasoning tasks highlighted by our stakeholders in Chapter 2 as being most likely to be facilitated by advanced mathematical study. In most cases, we found the pattern of results that the TFD would predict: the mathematics group's responses were closer to the norms of classical logic than the comparison group's. This was particularly the case with the tasks that involved abstract content, and especially the abstract conditional inference task.

However, as we discussed in Chapter 2, cross-sectional evidence like this should not be considered very persuasive evidence for the TFD. While the TFD would have been shaken by a finding that there is no difference in reasoning between mathematics and comparison students, the fact that there is leaves open two reasonable hypotheses: the TFD and what we have been calling the filtering hypothesis. One of the stakeholders we interviewed referred to this as the 'chicken and egg' situation: perhaps those who already reason differently are more likely to be filtered into studying advanced mathematics.

In this chapter, we aim to investigate this hypothesis directly. Three studies are reported. In the first, we tracked the reasoning development of groups of UK-based post-compulsory mathematics students, and compared it with the development of UK-based post-compulsory English literature students. The second study was a follow-up, which involved investigating first year undergraduate students' conditional reasoning. In the third study reported in this chapter we repeated the first study in Cyprus, an educational system where studying mathematics is compulsory until the age of 18.

4.1   Does Studying A Level Mathematics Develop Conditional Reasoning?

Our main goal in this study was simply to investigate whether or not the conditional reasoning behaviour of students changes over the course of their A level studies.[1] Recall that 16-year-old students in England and Wales typically choose to study three or four subjects at A level. A levels are two-year courses that are the primary means by which British universities select students. At the time we conducted the study, the first year of the A level programme was a standalone qualification known as an AS level. So students could, after their first year, choose to progress with the full A level, or leave with an AS level qualification.

We recruited 124 16-year-old students to take part. Of these 77 had opted to study AS level mathematics, and 47 had opted to study AS level English literature and not mathematics. We asked our participants to take a reasoning test at two points: close to the beginning of their AS level studies, and close to the end of their first year of studies.

As noted in Chapter 2, longitudinal quasi-experimental designs that seek to investigate the TFD can be strengthened by taking measurements of factors that might predict the extent to which students' reasoning skills might develop regardless of their mathematical study. The work of Keith Stanovich suggests two sensible factors to focus on: students' cognitive capacity as indexed by classic intelligence tests, and their thinking dispositions, as indexed by self-report measures and the cognitive reflection test. Both of these seem plausible candidates for variables that might be predictive of reasoning gains during education.

As our measure of cognitive capacity, we used an 18-item subset of Raven's Advanced Progressive Matrices, the same standard intelligence test discussed in Chapter 3. Participants were given 15 minutes to solve as many of the Raven's items as possible. The 18 items were chosen based on prior research.[2]

We used two measures of thinking disposition. The first was a self-report scale called the Need for Cognition (NFC) scale.[3] This requires participants to state the extent to which they agree with 18 statements such as "I would prefer complex to simple problems" and "The notion

---

[1] This study was first reported in *PLOS ONE* (Attridge & Inglis, 2013).

[2] Stanovich & Cunningham, 1992.

[3] Cacioppo, Petty, Feinstein, & Jarvis, 1996.

of thinking abstractly is appealing to me". Our second measure was the cognitive reflection test, shown in Appendix A, Figure A.3. To disguise the 'trick' nature of these questions, which was a particular concern given that we would ask participants to solve the problems at the beginning and at the end of the year, we intermixed the three CRT items with three 'non-trick' mathematical word problems (in an earlier pilot study we demonstrated that this did not influence participants' responses to the CRT).

Our outcome measure was a 32-item version of the abstract conditional inference task. This consisted of the 16 items used in Chapter 3, together with another 16 where implicit negations replaced explicit negations (i.e. where a statement like "the number is not 3" was replaced by something like "the number is 7"). The problems were given to each participant in a random order, and the full set is given in Appendix B.

In addition, we took a measure of our participants' prior academic attainment (their previous year's examination results), and in order to check that the mathematics group did in fact learn mathematics during their year of study we also asked both groups to answer a short mathematics test made up of questions taken from various different standardised tests.

The students took part in the study during test conditions in groups during free periods in the school/college day. All the tasks were presented together in a single booklet. Participants all did the timed Raven's matrices section first, but the order of the remaining sections were counterbalanced between participants; that is to say that they all received the sections in different orders. The order of questions within the conditional inference and CRT and need for cognition sections were randomised for each participant.

As noted above, participants in the mathematics group were all studying the first year of A level mathematics. Although there are three different versions of this course available to students in England, all have similar content. Among other topics, the syllabus contained sections on algebra, geometry, calculus, trigonometry, probability, mathematical modelling, kinematics and forces. Most importantly, students were not taught any proof-based mathematics, nor were they taught the definition of the conditional statement. As discussed earlier, we formally established this by, as well as inspecting the syllabus, conducting an analysis of every first year A level mathematics examination between 2009 and 2011. Of 929 questions set, only one contained an explicit "if...then" sentence, and there were no mentions at all of the terms "modus ponens", "modus tollens" or "conditional".

As is normal in longitudinal studies, some of the students who took part at Time 1 did not take part at Time 2. There are many reasons why this might happen: the students may have moved schools, they could have been ill on the day the study took place, or they might have changed the portfolio of subjects they were studying. We checked to see whether those who dropped out of the study were systematically different to those who did not. Happily, there were no significant differences on any of the measures taken at Time 1 between those who dropped out and those who took part at Time 2, so we have no reason to suppose that this dropout was a substantial problem. In total, 82 participants took part at both time points and were included in our analysis.

We found that at Time 1 our mathematics group had significantly higher scores on Raven's, our measure of cognitive capacity,[4] and on the cognitive reflection test, one of our measures of thinking disposition.[5] There was also a marginal difference between the two groups in terms of prior academic achievement.[6] However, there was no group difference on the need for cognition scale, our other measure of thinking disposition. Given these findings, and our worries about the possibility of a Matthew effect, we controlled for Raven's and CRT scores, as well as prior attainment, in our main analyses reported below.

Happily we found that the mathematics group showed greater improvement on the mathematics test than did our comparison group of literature students, suggesting that they did indeed engage with and learn from their year of studying mathematics.[7]

In our first analysis we compared how endorsement rates for each of the four inferences we studied (MP, DA, AC and MT) changed for each group across the year. These are plotted in Figure 4.1. This graph requires some unpacking.[8] The first thing to note is that the responses of the two groups were very similar at Time 1: we found no significant differences between the two groups' endorsement rates for any of the four inferences.[9]

---

[4] $t(79) = 3.38, p = 0.001$.

[5] $t(79) = 4.79, p < 0.001$.

[6] $t(122) = 3.89, p = 0.089$.

[7] $F(1, 79) = 46.324, p < 0.001$.

[8] A $2 \times 4 \times 2$ ANOVA with two within-subjects factors: time (start and end of the year) and inference type (MP, DA, AC, MT), and one between-subjects factor: group (mathematics and literature) revealed a significant three-way interaction $F(3, 228) = 7.476, p < 0.001, \eta_p^2 = 0.090$, which remained significant when controlling for Time 1 Raven's, CRT and prior attainment scores, $F(3, 216) = 5.103, p = 0.002, \eta_p^2 = 0.066$.

[9] All $ps > 0.2$.

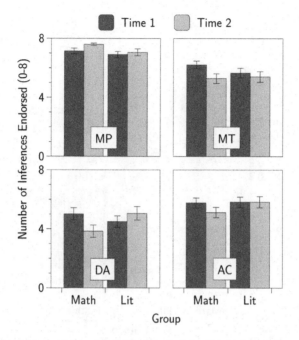

Figure 4.1 The mean number (max 8) of each inference endorsed by each group at Times 1 and 2. Error bars show ±1 standard error of the mean.

Secondly, both groups, as the literature would predict, had very high levels of endorsement of the valid MP inference. Thirdly, the literature group did not change their behaviour at all over the course of the year: there were no significant changes between their scores at Times 1 and 2.[10]

There were, however, substantial changes in the reasoning patterns of the mathematics students in the study. Compared to Time 1, the mathematics students at Time 2 rejected more of the invalid DA and AC inferences.[11] In other words, studying A level mathematics appeared to be associated with a greater tendency to reject invalid conditional inferences. However, somewhat surprisingly, the mathematics group also rejected more of the valid MT inferences at Time 2 compared to Time 1.[12]

---

[10]Although there was a marginally significant increase in the number of DA inferences endorsed by the literature group, $t(34) = 1.795, p = 0.082, d = 0.309$.

[11]$t(42) = 3.978, p < 0.001, d = -0.607$ and $t(42) = 3.060, p = 0.004, d = -0.468$, respectively.

[12]$t(42) = 2.877, p = 0.006, d = -0.446$.

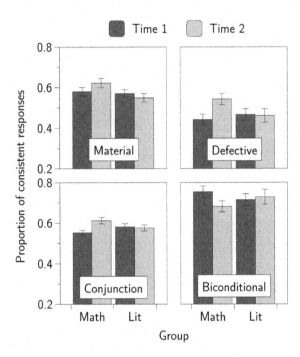

Figure 4.2   The mean proportion of responses consistent with each of the four interpretations of the conditional given by each group at Times 1 and 2. Error bars show ±1 standard error of the mean.

These responses appear most consistent with an increased tendency for the mathematics students to adopt a defective conditional interpretation (more MP inferences and fewer DA, AC and MT inferences were made at Time 2 compared to Time 1). To test for this, we calculated four indices for each participant (at each time point) giving the proportion of responses consistent with each of the four interpretations of the conditional discussed earlier in the book. These are shown in Figure 4.2.

Looking at the results this way indicates that the mathematics group became more material, whereas the literature group did not change.[13] Similarly, the mathematics group became less biconditional, whereas the literature group did not change.[14] The mathematics group became more

---

[13]Interaction: $F(1, 76) = 11.860, p = 0.001, \eta_p^2 = 0.135$ ($p = 0.007$ with covariates). Mathematics: $t(42) = 3.171, p = 0.003, d = 0.493$; literature: $p = 0.092$.

[14]Interaction: $F(1, 76) = 7.966, p = 0.006, \eta_p^2 = 0.095$, although this was only marginally significant when covariates were included, $F(1, 72) = 3.697, p = 0.058, \eta_p^2 = 0.049$. Mathematics: $t(42) = 3.323, p = 0.002, d = -0.508$; literature: $p = 0.500$.

defective, whereas the literature group did not change,[15] and finally, the mathematics group became more conjunctive, whereas the literature group did not change.[16]

Comparing the effect sizes of the material, biconditional, defective and conjunction analyses (ds = 0.493, −0.508, 0.880, and 0.693, respectively) suggests that the change in the mathematics group is best understood as an increased tendency to adopt the defective interpretation of the conditional. In other words, that over time the mathematics group became more likely to endorse the MP inference, but less likely to endorse the DA, AC and MT inferences. Strangely, this means that our mathematics group exhibited both *more* normative behaviour over time, and also *less* normative behaviour over time. They seemed to become more adept at rejecting the DA and AC fallacies, but this development appeared to be at the cost of also becoming less likely to endorse the valid MT inference.

We also considered whether these changes could be accounted for by either existing group differences in cognitive capacity or thinking disposition, or changes in these characteristics over the course of the year.[17] We found evidence for neither of these suggestions: only whether or not the student studied mathematics was a reliable predictor of changes in conditional reasoning behaviour. This seems to suggest that the changes we observed in the mathematics group are most likely to be related to experiences gained in their mathematical study, not to more general changes in cognitive capacity or thinking disposition.

Overall, the results of this longitudinal study are extremely encouraging for the TFD, but not overwhelmingly so. We found that studying A level mathematics seems to be associated with changes in conditional reasoning, even though the course itself contains no explicit tuition on the topic. However, contrary to the expectations of most TFD theorists, we found that the mathematics students developed in a not-entirely-normative fashion. Although they became less likely to endorse invalid inferences, they also became less likely to endorse the valid modus tollens inference. We have characterised this as adopting a more defective interpretation of the conditional although, as noted in earlier chapters, this language perhaps

---

[15]Interaction: $F(1, 76)$ = 17.651, $p < 0.001$, η_p^2 = 0.188 (p = 0.002 with covariates). Mathematics: $t(42)$ = 5.756, $p < 0.001$, d = 0.880; comparison: p = 0.767.

[16]Interaction $F(1, 76)$ = 8.525, p = 0.005, η_p^2 = 0.101 (p = 0.014 with covariates). Mathematics: $t(42)$ = 3.534, p = 0.001, d = 0.548; comparison: p = 0.693.

[17]A detailed report of the regression analyses involved in these tests is given in our *PLOS ONE* paper; Attridge & Inglis, 2013.

unhelpfully suggests that this way of thinking about the conditional is flawed. To be clear, this is not our view, and we discuss this issue further in Chapter 5.

When we first observed this MT result we were somewhat surprised. It seems inconsistent with both the predictions made by enthusiastic TFD supporters and their opponents. In view of this, we decided to conduct a follow up study to see whether we could detect this defective interpretation in other mathematical students, using a different task.

Before reporting this follow-up study, it is worth mentioning that we also investigated whether or not there were between-groups differences in terms of the negative conclusion and implicit negation effects. Readers will recall that in our discussion of the conditional inference task we noted that reasoners are more likely to endorse inferences that have negative conclusions than those with positive conclusions. Similarly, they are more likely to endorse inferences with explicit negations in their premises than those with implicit negations (i.e. "not 7" rather than "6"). To investigate this, we calculated measures of both these effects for each participant. A participant's negative conclusion index (NCI) was simply the difference between the number of inferences they endorsed with negative conclusions and the number they endorsed with positive conclusions; and their implicit negation index (INI) was the analogous difference between the number of inferences endorsed with implicit and explicit negative premises. Although both groups showed the typical negative conclusion and implicit negation effects, we found no evidence of different patterns of development for the two groups.[18]

In some sense these results are unsurprising given that we could not account for the overall observed changes in reasoning behaviour using our measures of cognitive capacity or thinking disposition. Recall that our favoured account of the negative conclusion effect concerns a Type 2 based double negation effect: endorsing a positive conclusion sometimes involves an extra step (for example deducing 'R' from 'not not R'), which increases the cognitive demand of the task. Given that we could not account for the overall group differences in development using changes in cognitive capacity, as indexed by our Raven's measure, it is consistent that there

---

[18]We analysed both indices using a 2×2 ANCOVA with one within-subjects factor (Time: Time 1, Time 2), one between-subjects factor (group: mathematics, English literature), and three covariates (prior academic achievement, Time 1 Raven's scores, and Time 1 CRT scores). Neither analysis showed a significant time by group interaction effect. NCI: $F(1, 72) = 3.60, p = 0.062, \eta_p^2 = 0.05$; INI: $F < 1$.

was also no between-groups difference in development with regards the negative conclusion effect. Also recall that we accounted for the implicit negation effect by appealing to Type 1 heuristic processes (the premise 'not 3' is more obviously relevant to a rule containing the symbol '3' than is the premise '6'). If changes in the prevalence of the implicit negation effect were behind the emergence in overall group differences in conditional reasoning behaviour, we might have expected there to have been a relationship between our measure of thinking disposition (the CRT) and conditional inference development. But we found no such association. In summary, it seems that our primary finding — that the mathematics groups developed their conditional reasoning behaviour over the course of a year of mathematical study — does not seem to be related to either the negative conclusion effect or the implicit negation effect.

4.2 Do Mathematics Students Respond Defectively to the Truth Table Task?

In the study reported in the last section we found that studying mathematics A level appears to be associated with changes in conditional reasoning behaviour, but towards the non-normative defective conditional interpretation rather than the normative material conditional interpretation. To investigate this further we compared two groups of first year undergraduates on a different task that involves conditional reasoning: the truth table task.

The conditional inference task, used in the previous two studies, asks participants to decide whether or not a given deduction can be legitimately made from a premise and a conditional. In contrast, the truth table task presents participants with a conditional rule and a situation which either conforms to the rule, contradicts the rule or is irrelevant to the rule. Here is how the task is presented to participants:

> This problem relates to cards which have a capital letter on their left hand side and a single digit number on the right.
> You will be given a rule together with a picture of a card to which the rule applies. Your task is to determine whether the card conforms to the rule, contradicts the rule, or is irrelevant to the rule.
> *Rule:* If the letter is A then the number is 1.
> Card: A 1
>   ○   conforms to the rule
>   ○   contradicts the rule
>   ○   irrelevant to the rule

Very clearly the card $\boxed{\text{A 1}}$ is consistent with the rule 'if A then 1'. But what about the card-rule pair $\boxed{\text{G 4}}$ and 'if B then 3'? This latter problem allows us to distinguish between the defective and material interpretations of the conditional. According to the material interpretation, the rule 'if p then q' is true whenever not-q is true. So because 4 is an instance of not-3, the card $\boxed{\text{G 4}}$ conforms to the rule 'if B then 3'. In contrast, under the defective interpretation, the card $\boxed{\text{G 4}}$ is irrelevant to the rule because G is an instance of not-B. By asking about a variety of different card and rule pairs, it is possible to investigate the extent to which a participant behaves in a manner consistent with the material, defective and biconditional interpretations of the conditional.

If the results from the longitudinal study reported in the last section indicate that experience of studying A level mathematics is associated with a development of a defective interpretation of the conditional, then we would expect to see a group difference in conditional reasoning behaviour on the truth table task. Specifically, we would expect a sample of first year undergraduate mathematics students to show a more defective pattern of responses to the truth table task than a sample of first year undergraduate arts students. To investigate this we asked two such groups to take the task. We recruited 33 first year mathematics undergraduates and 21 first year arts undergraduates from a well-respected UK university to take part, and asked them to tackle the truth table task, along with a short measure of cognitive capacity, the AH5 intelligence test.[19]

Each participant responded to 32 truth table problems of the form shown above. Given the rule 'if A then 1' the card $\boxed{\text{A 1}}$ can be thought of as a TT (true-true) instance, as it shows a true antecedent (A) and a true consequent (1). Similarly, the card $\boxed{\text{G 1}}$ is an instance of FT (false-true). For each of the four possible rules formed, as in the conditional inference task, by rotating the presence of negations in the rule, we asked participants about two TT, TF, FT and FF cards. The complete set of problems is given in Appendix F.

The proportion of responses consistent with the material, defective and biconditional interpretations of the conditional are shown in Figure 4.3. While there were no significant group differences in terms of the material or biconditional interpretations,[20] there were significant group differences

---

[19]Heim, 1968.

[20]Material: $t(36.5) = 0.125, p = 0.901$; Biconditional: $t(29.9) = 1.333, p = 0.193$.

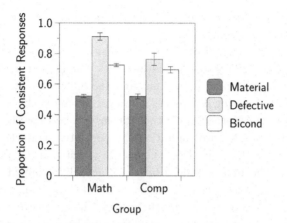

Figure 4.3   The proportion of responses from each group consistent with each interpretation of the conditional. Error bars show ±1 standard error of the mean.

on the defective conditional: a very high proportion of responses from the mathematics group, 91%, were consistent with the defective conditional, compared to only 76% in the comparison group, a difference of nearly 1 standard deviation.[21] As before, we found that our mathematics group scored higher on our measure of cognitive capacity,[22] but this could not account for the difference we found between the two groups' defective responses: when we controlled for AH5 scores, the group difference remained.[23]

So, the results of this study seem highly consistent with those from the longitudinal study of A level students. There we found that our mathematics and comparison groups started off by exhibiting similar behaviour on the conditional inference task, but by the end of their first year of study they had diverged substantially. Perhaps surprisingly, the mathematics group's change in behaviour could best be characterised by becoming less biconditional and more defective. Here we used a different task to investigate the differences between a group of first year undergraduate mathematics students, who had all studied A level mathematics or some equivalent course, and a comparison group of undergraduate arts students. We found consistent results: there were significant between-groups differences in behaviour on the truth table task, and this was largely because

---

[21]$t(34.2) = 3.197, p = 0.003, d = 0.951.$
[22]$t(50.1) = 2.532, p = 0.015.$
[23]$F(1, 51) = 5.996, p = 0.018.$

the mathematics group adopted a more defective interpretation of the conditional than the comparison group.

Overall then, our findings from these two studies should encourage supporters of the TFD: studying mathematics does appear to be associated with changes in reasoning behaviour, and these changes cannot easily be explained by pre-existing differences. However, we found that the reasoning development in our mathematics students was not best characterised by an increasing tendency towards adopting the normative material conditional interpretation. Rather, the frequency with which mathematics students endorsed the valid modus tollens inference actually dropped during their year of studies. Why this might be deserves more consideration, and we return to the issue later in the book. First, however, we consider a further issue raised by these studies: whether the extent to which mathematics students develop their reasoning is associated with the amount of mathematics they study. To do this we ran a version of our longitudinal study in Cyprus.

4.3   Post-compulsory Mathematical Study and Reasoning Development in Cyprus

Cyprus is an interesting country for several reasons.[24] First, it is in some sense a more typical country than England or Wales, as students are required to study mathematics until the age of 18. As noted earlier, allowing students to drop mathematics at the age of 16 is highly unusual.[25] In Cyprus 16-year-old students choose between studying 'high intensity' and 'low intensity' mathematics. Both programmes are studied over two years: the low-intensity mathematics syllabus involves three 45-minute sessions per week in the first year and two in the second year. The high-intensity syllabus involves seven 45-minute sessions per week in the first year and six in the second year.

Of particular interest is the large Euclidean geometry component of the high-intensity syllabus, which is quite different to the A level syllabus studied by the students in our earlier longitudinal study. Deductive geometry is perhaps the topic which we would expect to be most associated with the development of reasoning skills. For instance, the Cambridge historian of

---

[24]This study was first reported in *Research in Mathematics Education* (Attridge, Doritou, & Inglis, 2015).
[25]Hodgen *et al.*, 2010.

mathematics Piers Bursill-Hall noted that geometry has traditionally been regarded as "the finest way to train the mind, the most perfect training in reasoning and clear thinking", and that it is "the ultimate exercise in clear, logical thinking and reasoning".[26]

So, together with our collaborator Maria Doritou, we recruited 188 Cypriot 16-year-olds to take part in a two-year longitudinal study which tracked their reasoning development using a similar design we used in the longitudinal study in England. We first translated all our items into Greek, and then back translated them to ensure comparability. This task is not as straightforward as it seems, as there are (at least) two different linguistic forms of "if p then q" in Greek:

- Αν p τότε q.
- Εαν p τότε q.

The latter, the 'Εαν' form, is considered to be mildly more formal. One Greek scholar we spoke to suggested that the situation might be analogous to the difference between speaking English in a Cockney accent compared to standard received pronunciation. While the 'Εαν' form might be more authoritative, we opted to use 'Αν', as it seemed to be more common in the selection of Greek-language mathematics textbooks we had access to. Thus a full problem looked something like this:

Αν το γράμμα είναι το Α τότε ο αριθμός είναι το 3.
Το γράμμα είναι το Α.
Συμπέρασμα: Ο αριθμός είναι το 3.
○ ΝΑΙ
○ ΟΧΙ

All of our 188 participants studied in the same school, and around 40% opted to study high-intensity mathematics. The school used a copy of *Euclidean Geometry General Lyceum A and B* as their textbook for the high-intensity classes, which gives us a sense of the style of mathematics taught.[27] An earlier edition of this textbook was described as following a "traditional Euclidean deductive style, in which the definitions and the enunciations of theorems are not motivated but imposed upon the reader" and that it "emphasises tasks that require a formal proof of some

---

[26]Bursill-Hall, 2002, pp. 28 & 30.
[27]Argyropoulos *et al.*, 2010.

mathematical fact"[28] A full description of the content of the high- and low-intensity courses is given in Appendix G.

As before, the students took part in the study under test conditions in groups during free periods in the school day. All the tasks were presented together in a single booklet. Participants all did the timed Raven's matrices section first, but the order of the remaining sections were counterbalanced between participants, and the order of questions within the conditional inference and CRT and need for cognition sections were randomised for each participant.

Of the 188 participants who started the study (74 in the high-intensity group and 114 in the low-intensity group), a large majority returned at the second and third time points (184 and 180, respectively). However, not all participants completed all tasks at each time point: 107 (44 high-intensity and 63 low-intensity) completed the conditional inference task at all three time points. Those who attended but did not complete tasks either missed them out or left them incomplete. This level of drop-out is potentially a threat to the validity of our findings, so we investigated whether or not those students who completed all tasks had different levels of performance at Time 1 from those who didn't. Happily, we found no significant differences on any of our measures at Time 1.[29] Therefore, our analysis proceeded with the 107 participants who completed the conditional inference task at all time points, as well as the sections on Raven's matrices and the CRT.

Given our worries about a possible Matthew effect, we first investigated whether we could predict participants' changes in conditional inference scores using our Raven's and CRT measures. As in the English study, we found group differences on both these measures at the first time point: compared to the low-intensity group, the high-intensity students had higher Raven's and CRT scores[30] Therefore, we controlled for these variables in our main analyses.

The number of inferences of each type endorsed by each group at the three time points is shown in Figure 4.4. Several things stand out from this graph. First, as in the English study, there were no significant differences between the groups at the start of their post-compulsory study: at Time 1, both groups performed similarly. Second, again like in the English study,

---

[28] Mouzakitis, 2006, pp. 15 & 17.

[29] All $ps > 0.1$.

[30] Raven's: $t(180.1) = 5.621, p < 0.001, d = 0.788$. CRT: $t(84.3) = 2.060, p = 0.043, d = 0.379$.

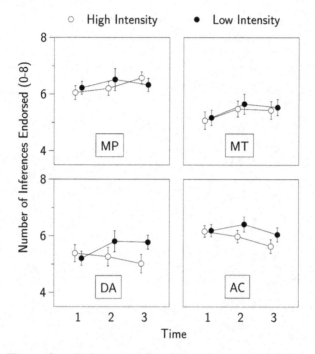

Figure 4.4 The number of inferences endorsed by each group at each time point. Error bars show ±1 standard error of the mean.

the two groups had diverged by the end of the study. Third, this divergence again seemed to be largely related to the two invalid inferences: denial of the antecedent and affirmation of the consequent. In both cases, the high-intensity group endorsed fewer of these inferences over time. Fourth, in contrast to the mathematics students in our English study, we found no evidence that the high-intensity Cypriot students endorsed fewer modus tollens inferences over time.

When we calculated the proportion of responses consistent with the material and defective interpretations of the conditional, shown in Figure 4.5, we found that the reasoning development of the two groups did differ significantly.[31] However, whereas in the English study this difference was

---

[31]On a 3 (time) by 2 (group) ANCOVA, with the material conditional index as the dependent variable, and Raven's and CRT as covariates, the time by group interaction effect was significant, $F(2, 158) = 6.396, p = 0.002, \eta_p^2 = 0.075$. However, the equivalent effect for the defective conditional index did not reach significance, $p = 0.589$, largely because the two groups had similar patterns of development on the MT problems.

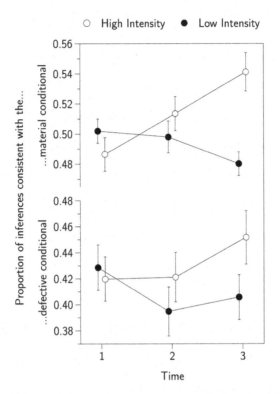

Figure 4.5   The proportion of responses consistent with the material and defective interpretations of the conditional. Error bars show ±1 standard error of the mean.

best characterised by a change in the mathematics group towards the defective conditional, here we found that the change in the Cypriot high-intensity students was best described by a shift towards the material conditional. However, both studies were consistent in that the major group differences emerged in the invalid inferences: the difference was that the Cypriot high-intensity mathematics students did not consistently change their behaviour on MT problems.

Can we say more about how the different curricula influence students' responses to the invalid inferences? In Figure 4.6, we have plotted the change in the number of invalid inferences endorsed by each group in their first year of post-16 study on the y-axis, with the approximate proportion of the curriculum each group devoted to mathematics on the y-axis (this is approximate, as not all the students in the UK study studied for the same numbers of hours per week). So, for instance, on average the AS English

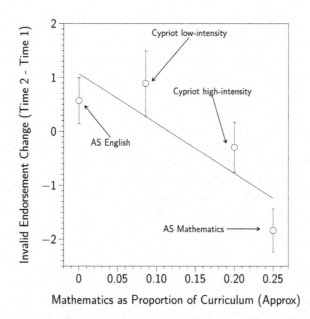

Figure 4.6   The mean change in invalid inferences over the course of a year, by the approximate proportion of the curriculum devoted to mathematics. Error bars show ±1 standard error of the mean.

group endorsed around 0.5 more invalid inferences at Time 2 compared to Time 1.[32]

With the important caveat that we only have four datapoints on this graph, we can see that there does appear to be a link between the proportion of the curriculum a student spends on mathematics, and the tendency to endorse fewer invalid conditional inferences. This relationship can be quantified using a Spearman correlation coefficient, $r = -0.275$, a highly significant association.[33]

One surprising finding from our Cypriot data was that the participants — in both groups — endorsed the straightforward modus ponens inference much less than their English counterparts in the earlier study. Recall that the modus ponens inference merely consists of deducing q from p and 'if p then q'. Endorsement rates for this inference are typically extremely

---

[32] Also consistent with the A level study, we found no evidence that these differences were related to the negative conclusion effect or implicit negation effect. When we calculated NCI and INI indices and conducted the appropriate analyses, we found no evidence of time by group interactions for either inference, both $ps > 0.1$.
[33] $p < 0.001$.

high.[34] Among our English students the inference was endorsed around 90% of the time, but the equivalent figure for Cypriots was only around 78%. One tentative explanation for this curious discrepancy relates to our earlier discussion of the difference between the less formal 'Αν' version of 'if' and the more formal 'Εαν' version. Some Greek speakers suggest that 'Εαν' carries "more authority" than the 'Αν' form; perhaps the salience of the modus ponens inference could be increased by increasing the formality with which the conditional in question is expressed. This suggestion merits further investigation, but to do so here would take us too far away from our primary purpose of considering the TFD.

Overall, the results of our study in Cyprus are, like its counterpart which took place in England, somewhat encouraging for the TFD. In the next section, we summarise what the studies reported in this chapter tell us.

4.4   Summary

Our goal in this chapter was to investigate whether the group differences in reasoning behaviour we found in Chapter 3 — mathematics students appeared to reason about conditionals differently to non-mathematics students — might be explained by the TFD. In other words, did we find these differences *because* the mathematics students had studied mathematics? Or, alternatively, might these differences be better explained by what we have been calling the filtering hypothesis? Are people who reason in a particular way more likely to choose to study post-compulsory mathematics?

We investigated these questions through two longitudinal studies. In both cases we tracked students at the beginning of their post-16 studies. In England, students at this age can drop mathematics entirely, whereas in Cyprus they must choose to study either high- or low-intensity mathematics. In both our English and Cypriot studies, we found that groups of students who made different mathematical study choices at sixteen started with similar conditional reasoning behaviours. However, those groups which chose to study a greater amount of mathematics developed their conditional reasoning in a way that the groups who chose to study no mathematics, or only low-intensity mathematics, did not.

These results seem more consistent with the TFD than the filtering hypothesis. If pre-existing differences were the cause of the group differences we found in Chapter 3 then we would have expected to be able to detect

---

[34]Evans & Over, 2004.

group differences at Time 1 in both our longitudinal studies. In neither case were we able to.

However, the results are not completely compatible with the TFD, or at least not the version of the TFD put forward by the stakeholders we interviewed in Chapter 1. The view taken by our interviewees was that studying mathematics would develop a normative understanding of conditional statements. In other words that they would become more likely to interpret 'if ... then' statements as material conditionals.

In fact this is not what we found. In both the English and Cypriot studies we found that the largest change in the mathematics groups was towards becoming more likely to reject the invalid denial of the antecedent and affirmation of the consequent inferences. The Cypriot mathematics students did not change a great deal in their behaviour on the valid modus tollens inference, whereas the English mathematics group actually became less likely to endorse this inference. This result surprised us, so we ran a second study which investigated whether we could detect the same trend in a group of first year undergraduate students. A consistent pattern emerged.

There are several ways of characterising these changes. One way is to say that, while our mathematics groups did become more normative, it might be more accurate to describe the change as away from the biconditional interpretation of 'if p then q' and towards the defective interpretation. Another way would be to simply say that studying mathematics was associated with becoming more sceptical or critical of inferences. While this change led to a reduced likelihood of accepting invalid inferences it also, at least in the English group of mathematics students, led to a reduced likelihood of accepting the valid MT inference as well.

Two interesting questions arise from this finding that A level mathematics students seem to become less likely to endorse the MT inference over time. First, is it a concern? Is the modus tollens inference an important part of mathematics which we hope that our students should be able to successfully engage with? Second, what happens when students engage in more advanced mathematical study? Do they continue to reject the MT inference, or does the developmental pattern reverse? It is these questions to which we turn the next chapter.

Chapter 5

The Modus Tollens Inference
and Mathematics

In Chapter 4, we saw that our group of post-compulsory A level mathematics students became more likely to reject the valid modus tollens inference over time. Our goal in this chapter is to explore this finding further, in two directions. First, we ask whether this is an educational problem: is it important that mathematics students be able to make MT deductions? Second, we ask how the propensity to make the MT inference changes with more advanced mathematical study. In particular, we report a longitudinal study of undergraduate mathematics students, similar in form to those reported in Chapter 4.

5.1 Modus Tollens and Advanced Mathematics

The issue of how mathematics students should understand conditionals like 'if p then q' has received some attention in the mathematics literature. Like many mathematical concepts, the formal definition of 'if p then q' — by which we mean the material conditional — does not always match students' intuitive understandings of the concept. One way of understanding this discrepancy is to use the notion of a concept image, an idea first proposed by the influential mathematics education researchers David Tall and Shlomo Vinner.[1]

The idea of a concept image is simple yet powerful: an individual's concept image is the total cognitive structure that they associate with the given concept. This is a potentially huge collection of properties, pictures or processes that are connected in some way with the particular concept. A concept image might be completely informal, and it might not be coherent. Parts of the concept image might not agree with other parts. This won't

---

[1] Tall & Vinner, 1981.

cause problems, however, unless the conflicting parts of the concept image are evoked simultaneously. Tall and Vinner contrasted the notion of a concept image with that of a concept definition, which they described as a form of words used to specify the concept.

There are many classic examples of students having concept images which are not consistent with the formal concept definition shared by the mathematical community. For instance, it has often been observed that students struggle with the idea that a sequence of real numbers can exceed it's limit.[2] Whereas the natural language word "limit" implies some kind of barrier which cannot be crossed, the formal concept definition carries no such implication.

However, concept definitions are not always primary over concept images. The notion of a curve provides a good example. We all have a clear intuitive understanding of what a curve is: when pushed to offer a definition or characterisation we would probably respond with some discussion of a smooth line that divides the plane in two (at least locally). In other words, we all have a clear concept image of a curve. But what is the concept definition? Perhaps surprisingly, there have been many attempts to provide definitions, but it transpires that none successfully captures our concept image.

For instance, defining a curve to be the image of any continuous function which maps the interval $[0, 1]$ to the plane, results in the surprising and unpalatable conclusion that the complete unit square is a curve.[3] This is clearly unacceptable: squares fall well outside our concept image. An alternative approach is to use topological notions of dimensionality, compactness and connectedness. We can exclude squares from the family of curves by defining any one-dimensional, compact and connected subspace of the plane to be a curve.[4] Unfortunately, this leads to a variety of strange objects becoming curves. A nice example is the three-sided curve of G. T. Whyburn. He constructed a peculiar curve which has a section where every point borders three separate parts of the plane rather than two.[5] Perhaps even more puzzling is the curve described by the American mathematician William Fogg Osgood. He found a curve which was one

---

[2]Monaghan, 1991.

[3]For a fascinating discovery of so-called space filling curves, see Sagan (1994).

[4]Readers interested in understanding these topological concepts are referred to Armstrong, 1983.

[5]Whyburn, 1942.

dimensional, which didn't cross itself anywhere, which had finite length, but which nevertheless covered a positive area.[6]

The point of these examples is to show that the concept definition is not always primary to the concept image. If we have a strong intuition that a particular object should not fall within a given category or behave in a given way, then the fact that it meets our definition might constitute a criticism of that definition, rather than a critique of our understanding. Is this what is happening with the material interpretation of the conditional? Maybe the fact that it fails to capture our concept image of 'if p then q' means that the material interpretation should simply be abandoned as a concept definition, and that it should be replaced by the defective interpretation?

In fact this position has been adopted by some influential mathematics educators. Celia Hoyles and Dietmar Küchemann argued exactly this, at least in the context of school mathematics. They wrote that

> when studying reasoning in school mathematics, the [defective conditional] is a more appropriate interpretation of logical implication than the [material conditional], since in school mathematics, students have to appreciate the consequence of an implication when the antecedent is taken to be true.[7]

Their argument was that the crucial aspect of conditionals that we need students to understand is that the consequent has to be true whenever the antecedent is. This is successfully captured by the defective understanding, so the material interpretation is superfluous.

But other mathematics educators have disagreed. Viviane Durand-Guerrier offered two reasons to reject Hoyles and Küchemann's argument.[8] First, she offered an example of a mathematical definition which she claimed required the material conditional to understand, namely that of a diagonal matrix. This is any square matrix which has zeros everywhere except on its diagonal (which may or may not have zeros on). Examples would be:

$$\begin{pmatrix} 1 & 0 & 0 \\ 0 & -3 & 0 \\ 0 & 0 & 2 \end{pmatrix} \quad \text{or} \quad \begin{pmatrix} -\frac{1}{3} & 0 & 0 & 0 \\ 0 & \pi & 0 & 0 \\ 0 & 0 & 15 & 0 \\ 0 & 0 & 0 & 0 \end{pmatrix}.$$

A typical concept definition for a diagonal matrix would be something like:

---

[6] Osgood, 1903.
[7] Hoyles & Küchemann, 2002, p. 196.
[8] Durand-Guerrier, 2003.

An $n \times n$ matrix $[a_{ij}]$ is diagonal if and only if for every i from 1 to n, and every j from 1 to n, if $i \neq j$ then $a_{ij} = 0$.

The crucial case here is that of $i = j$. The material conditional 'if $i \neq j$ then $a_{ij} = 0$' is true in this case regardless of whether $a_{ij} = 0$ or $a_{ij} \neq 0$, so the definition accurately captures our concept image of a diagonal matrix.

But is Durand-Guerrier really correct to say that understanding this definition requires a material interpretation of the conditional? It is certainly true that a biconditional interpretation would cause problems: interpreting the conditional in this manner would rule out the possibility of diagonal matrices where some of the diagonal entries are zero. But what about a defective interpretation? Someone adopting such an interpretation would think that the conditional 'if $i \neq j$ then $a_{ij} = 0$' was simply irrelevant in cases where $i = j$. In other words, they would think that cells off the diagonal had to be zero, and that the definition didn't say anything about the content of cells on the diagonal. So the defective interpretation of the conditional also seems to lead to a definition which captures the appropriate concept image.

Durand-Guerrier's second argument against Hoyles and Küchemann's position revolved around the modus tollens deduction. She argued that one cannot successfully draw MT inferences if one adopts a defective interpretation.[9] She might be overstating the case here: it does seem possible to draw the MT inference using MP together with an informal contradiction argument, an approach which could be accessible to the defective reasoner. For instance, given the premises 'if p then q' and not-q, one might suppose p, conclude q by MP, notice that q contradicts the assumption not-q, and so conclude that the supposition p was incorrect, concluding not-p. Nevertheless, it certainly seems plausible to suppose that this line of reasoning would require more working memory capacity than directly concluding not-p from 'if p then q' and not-q using MT.

So is there any merit to Durand-Guerrier's critique of Hoyles and Küchemann's argument? The issue can be investigated empirically. Essentially, Durand-Guerrier's argument boils down to saying that the MT inference is important for mathematical success, and that this is even more the case at more advanced levels of mathematics. Hoyles and Küchemann are arguing the reverse: that making the MT inference is largely irrelevant to mathematical success. To investigate this question we ran an experiment,

---

[9]Durand-Guerrier, 2003, p. 29.

in collaboration with our colleagues Lara Alcock, Toby Bailey and Pamela Docherty, in which we investigated how mathematical achievement is related to conditional inference behaviour.[10]

We recruited 112 first year undergraduate mathematics students to participate in the study. All were studying mathematics at a well-regarded UK university and had taken (or were taking) modules on linear algebra, problem solving and proving, and calculus. We measured conditional reasoning using the 16-item abstract version of the conditional inference task used in earlier chapters.[11] This contained four items for each of the MP, MT, DA and AC inferences.

We took two measures of mathematics achievement. The first was the mean examination scores for the students' calculus, linear algebra, and problem solving and proving modules. Each of these modules were assessed using traditional summative examinations that were worth at least 85% of the overall module score. Our second achievement measure focused on mathematical proof comprehension.

Proofs are fundamental to advanced mathematics, and a large proportion of undergraduate mathematics learning takes place by reading written proofs, perhaps in textbooks, perhaps on blackboards in lectures. Unfortunately, however, students find this to be an extremely difficult task. Many research studies have shown that students struggle to read proofs effectively.[12] Since proofs are essentially just logical arguments, it would seem plausible to hypothesise that a capacity for accurate logical reasoning would predict proof comprehension skills. But does an ability to reject the invalid MT inference predict proof comprehension skills? To find out, alongside our measure of mathematics examination achievement, we also included a measure of proof comprehension.

We first asked our participants to read a proof of a theorem in number theory that stated that the product of two distinct primes is not abundant. An abundant number is a positive integer n whose divisors add up to more than $2n$. So, for instance, 12 is abundant because the divisors of 12 are 1, 2, 3, 4, 6, and 12, and these sum to 28 which is greater than 24. The theorem asserts that when any two primes are multiplied together the result cannot be abundant. So, for instance, 3×7 is 21, and the divisors of 21 sum to 32, which is less than 2×21. After participants had read a proof of this

---

[10]Alcock, Bailey, Inglis, & Docherty, 2014.

[11]Adapted from that used by Evans *et al.*, 1995.

[12]e.g. Alcock, Hodds, Roy, & Inglis, 2015; Inglis & Alcock, 2012; Selden & Selden, 2003.

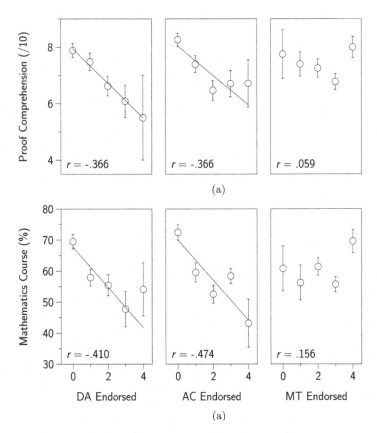

(a)

Figure 5.1   The relationship between endorsement rates and mean scores on the proof comprehension test (a) and formal examinations (b). Error bars show ±1 SE of the mean.

theorem, we asked them ten multiple choice questions about it. The full proof and associated comprehension questions are given in Appendix H.

The relationship between endorsement rates for the DA, AC and MT inferences and our two outcome measures are shown in Figure 5.1. There was not enough variance in endorsement rates for MP to conduct a similar analysis, as almost all participants endorsed almost all MP inferences.

The pattern of results shown in Figure 5.1 seems clear. While endorsement rates for the invalid DA and AC inferences were negatively correlated with success in both university mathematics exams and our proof comprehension test, there was no such relationship for the valid MT inference. In other words, the ability to reject invalid inferences seems to be predictive of university mathematics achievement, but the ability to endorse modus tollens seems not to be. That we found a similar pattern with both our measures of mathematics achievement should increase our confidence in our results.

The distinction between the material and defective interpretations revolves around the MT inference, but we found no evidence that this is related to success in undergraduate mathematics. This result seems inconsistent with Durand-Guerrier's view that the material conditional is necessary for successful engagement with advanced mathematics. Or at least if it is necessary, it must be for only a relatively small subset of advanced mathematics. One could hypothesise that direct access to MT is necessary to understand contradiction proofs, for instance. Further research would be needed to investigate such a proposal. But what we can say with confidence is that the propensity to reject invalid inferences is related to success in undergraduate mathematics, and it is precisely this ability which, according to the longitudinal studies reported in Chapter 4, appears to be developed by post-16 mathematical study.

5.2   Modus Tollens Development During Advanced Mathematical Study

In the last section, we saw that endorsing the modus tollens deduction is not associated with mathematical achievement, as measured by traditional examinations and by a proof comprehension test. But do undergraduates change their conditional reasoning behaviour over the course of their undergraduate studies? Of course, the TFD would predict that they do, and that their reasoning would become more normative over time. But we have already seen that studying mathematics between the ages of 16 and 18 was associated with improvements at rejecting the invalid inferences, not with an increased likelihood of endorsing the modus tollens inference. What happens when students study more abstract and rigorous university-level mathematics?

To investigate, we ran a version of the longitudinal conditional infer-ence studies reported in Chapter 4 with undergraduate participants. We recruited 65 single- or joint-honours mathematics undergraduates and 37 single- or joint-honours psychology undergraduates to serve as a comparison group. Like the A level students, these participants took part at two time points, at the beginning and end of their first year of undergraduate study. The measures we used were identical to those studies reported in Chapter 4, so rather than repeat ourselves let's jump straight to the results.

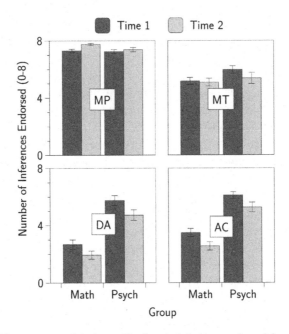

Figure 5.2   The mean number (max 8) of each inference endorsed by each group at Times 1 and 2. Error bars show ±1 standard error of the mean.

The mean number of each of the four main conditional inferences endorsed by each group at Times 1 and 2 are shown in Figure 5.2. We highlight several features of this graph. First, as before, and as expected, both groups endorsed MP nearly universally. Second, as we would expect given the results of the studies reported in Chapter 4, the mathematics group rejected more DA and AC inferences at Time 1 than did the psychology group. Everyone studying an undergraduate mathematics degree would have studied the equivalent of A level mathematics, so we can take the Time 1 group differences here to be some degree of confirmation of the results from Chapter 4. Third, again consistent with the results of the studies in Chapter 4, our mathematics group rejected more invalid inferences at Time 2 than they did at Time 1.[13] However, this time this was also true of our comparison group, made up of psychology students.[14]

---

[13]DA: $t(64) = 3.545, p = 0.001$. AC: $t(64) = 3.727, p < 0.001$.
[14]DA: $t(36) = 3.353, p = 0.002$. AC: $t(36) = 2.260, p = 0.030$.

Finally, neither group significantly changed their MT behaviour from Time 1 to Time 2, and for the mathematics group there was not even a trend.[15]

Overall then, these results seem consistent with both the studies reported in Chapter 4 and the study reported in the last section. We found that our mathematics undergraduates already performed differently to our control group at the start of their university studies, and that by the end of their first year they endorsed still fewer invalid DA and AC inferences. In the last section, we saw that the extent to which a student endorses MT doesn't seem to be related to their mathematics achievement, and the current results seem consistent with this view: we did not find that a year of undergraduate mathematics study changed the extent to which our undergraduates endorsed MT.

One further important lesson can be taken from this study. We found that the psychology group in this study became more normative over time. While this leaves open the possibility that both groups' development was simply the result of a general maturation effect, we do not think this is plausible. We have already seen in earlier studies that the English literature and low-intensity Cypriot groups did not show any such maturation effect. While this does not rule out the possibility of maturation having caused the psychologists' results entirely, it does render it rather improbable. The observation that our psychology group showed gains on the conditional inference task should remind us that although we have been focusing on mathematical study throughout this book, mathematics may not be the only academic subject that develops general thinking skills. Indeed, research conducted by Darrin Lehman and Richard Nisbett in the late 1980s suggested that psychology is particularly effective at developing skills of statistical and methodological reasoning.[16]

5.3 Summary

In this chapter, we have reported two studies which investigated the role of the modus tollens deduction in advanced mathematics. Earlier we found that somewhat surprisingly studying A level mathematics seems to be associated with making fewer MT deductions, contrary to what TFD advocates would have predicted. Here we investigated what this finding might mean: is MT important for advanced mathematics, as claimed by

---

[15]Mathematics: $t(64) = 0.341, p = 0.749$. Psychology: $t(36) = 1.721, p = 0.094$.
[16]Lehman, Lempert, & Nisbett, 1988; Lehman & Nisbett, 1990.

Viviane Durand-Guerrier? Or is rejecting invalid inferences the critical aspect of conditional inference? Both studies reported in this chapter support the latter position. We found that the frequency with which students endorse the MT inference is not associated with mathematical achievement, whereas the frequency with which they reject invalid DA and AC inferences is. Furthermore, we found that the extent to which mathematics undergraduates endorsed the MT inference did not change over their first year of study, whereas the extent to which they endorsed the DA and AC inferences reduced.

Overall then, the results of this chapter strongly suggest that the 'defective' understanding of the conditional is poorly named. Adopting a defective interpretation seems quite compatible with success in advanced mathematics. In terms of Tall and Vinner's concept image/concept definition framework, it may be that success in mathematics is compatible with students developing and using a concept image of the conditional that is not completely consistent with the formal concept definition.

Chapter 6

Conditional Inference Across the Mathematical Lifespan

In the last three chapters, we have presented converging evidence from a number of studies that studying advanced mathematics is associated with changes in conditional reasoning development. Although consistent results were found with various different reasoning tasks, our primary method of investigating these issues was to use the abstract conditional inference task, developed by Jonathan Evans.[1]

Using the same task consistently across different studies has the great advantage that it allows us to meaningfully compare the results. Our goal in this chapter is to summarise what we have learned so far about conditional inference development across what we might refer to as the post-compulsory 'mathematical lifespan'. Alongside the studies we reported in Chapters 4 and 5, we have included several other studies that we conducted, not reported in this book, that all used the same conditional inference task to assess mathematics students' conditional reasoning at different points in their mathematical studies. Because of international curricula differences, we have simplified matters by considering only UK-based students in this chapter.

There are a number of statistical problems involved in drawing conclusions by pooling data derived from studies with different designs. In particular, combining cross-sectional and longitudinal studies makes sample-to-population statistical inference difficult. For some of the data reported here we collected measures from participants at two time points (A level mathematics students at the start and end of their first year, for instance), whereas for some we collected data only once (third year

---

[1] Evans *et al.*, 1995.

mathematics undergraduates). Because of this, standard statistical methods cannot easily be applied. This limitation should be born in mind when considering the data reported in this chapter. With that caveat in mind, what patterns might we expect to see?

In Chapters 4 and 5, we put forward the suggestion that studying advanced mathematics is associated with an increased tendency to reject invalid conditional inferences, but also an increased tendency to reject the valid MT inference. This latter tendency seemed to be restricted to those studying A level mathematics: in our Cypriot study and our undergraduate study, no changes in MT inference endorsements were observed. If we assume that our participants in these studies come from the same population, but are merely at different stages of development, does a consistent pattern emerge?

In Figure 6.1, we have plotted results from four different studies. Two were reported in Chapters 4 and 5. The two extra studies involved students

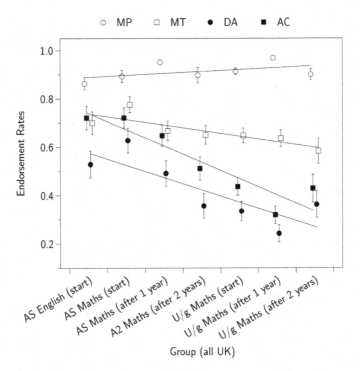

Figure 6.1   Endorsement rates for MP, DA, AC, and MT from various different studies involving participants at different education levels. Error bars show ±1 SE of the mean.

towards the end of their second year of A level study, and in their third and final year of a mathematics undergraduate degree at a respected UK university. On the graph, they are referred to as "A2 Maths" and "U/g Maths (after 2 years)", respectively. The x-axis is ordered in terms of mathematical experience. On the far left are the English literature students at the start of their post-compulsory studies (which, as discussed in Chapter 4, involved no mathematics at all). On the far right are final year undergraduates, who had at this point completed nearly five years of post-compulsory mathematics.

The lines of best fit for each inference type show a reasonably consistent pattern. Over the mathematical lifespan the tendency to endorse MP seems to increase slightly, albeit from a very high base. The frequency of endorsements of the invalid DA and AC inferences declines steeply from over 60% to around 30%. The tendency to endorse MT also shows a decline, albeit a much weaker one than the two invalid inferences.

The same data can be expressed by considering the extent to which participants' responses are consistent with the different interpretations of the conditional. Figure 6.2 shows the extent to which responses were consistent with the material, defective and biconditional interpretations. Again, the data show that the more experience someone has of studying advanced mathematics, the less they seem to respond in a manner consistent with the biconditional interpretation.

We have emphasised throughout that it is extremely difficult to provide strong evidence for the causal claim that is at the heart of the TFD. Does studying mathematics *cause* a change in reasoning behaviour? Without the ability to conduct a true experiment, where students would be randomly assigned to study advanced mathematics or not, it is impossible to give a definitive answer to this question. But, all is not lost: although we are not able to use random assignment to investigate the TFD, we can subject it to extremely stringent tests. In the last few chapters, we have reported a series of studies that did exactly that, and in each case we obtained data that are highly consistent with a version of the TFD. Before moving on, we briefly summarise the evidence reported so far.

In Chapter 3, we compared the syllogistic and conditional reasoning behaviour of students who had experienced advanced mathematical study and students who had not. If we had found no differences, the TFD would have been in trouble, but in fact we found substantial differences and the various covariates we collected alongside our primary reasoning measures were not sufficient to account for them. In other words, we found that

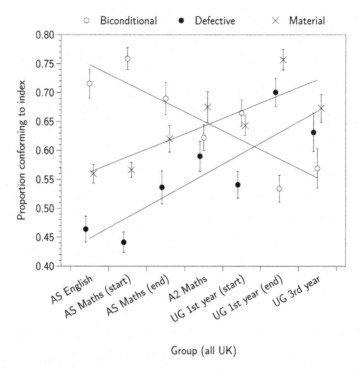

Figure 6.2   The proportion of responses consistent with the material, defective and biconditional interpretations of the conditional. Error bars show ±1 SE of the mean.

mathematicians and non-mathematicians do appear to reason differently, and that this cannot easily be explained by group differences in intelligence.

A stronger test of the TFD was to track students longitudinally. We did this across two studies reported in Chapter 4. In both cases we found that there were no differences in abstract conditional reasoning behaviour between our mathematics and comparison groups at age 16, but that by the end of their first year of studies the two groups had diverged. In our Cypriot study, we found that this divergence continued into the second year of post-compulsory study. In Chapter 5, we looked at the conditional reasoning behaviour of mathematics undergraduates, and found that their conditional reasoning behaviour also changed over the course of their first year of studies. In this chapter, we have brought all our work on abstract conditional inference together. It paints a consistent picture, which is entirely compatible with a version of the TFD that claims that one major benefit of studying advanced mathematics is that it reduces the likelihood of endorsing invalid conditional inferences.

Let us suppose that the causal version of this claim were true, and that studying advanced mathematics *causes* students to more often reject invalid conditional inferences. Why would exposure to advanced mathematics have this effect? It is this question that we turn to in the next chapter.

Chapter 7

Why Would Studying Advanced Mathematics Develop Conditional Reasoning?

Let us suppose that the evidence presented earlier in the book is sufficiently convincing for us to conclude that studying advanced mathematics probably does cause students to change the ways in which they interact with conditional 'if ... then' statements. Why would this be?

At least some aspects of this question can be investigated empirically. In particular, one unresolved issue concerns the extent of the changes in the mathematics students' conditional reasoning behaviour that we observed. One hypothesis is that mathematics students, as a result of their study, become increasingly familiar with forward deductions, often expressed using "if... then" statements. They learn that treating such statements as if they were "if and only if" statements leads to logical problems.

This hypothesis suggests that increased exposure to conditionals expressed in some linguistic form similar to 'if p then q' is the mechanism by which the development occurs. If this were the case, we might expect that the changes we observed with mathematics students' conditional reasoning would not generalise to problems where the conditional statements were expressed in a linguistic form that is rarely used in advanced mathematics.

Our question concerns the extent to which mathematics students' reasoning is generalised *within* abstract conditional inference. The conditional inference task used in the studies reported in earlier chapters consistently presented participants with conditional statements of the form 'if p then q', but there are other ways of expressing the same statement. For instance, a logically equivalent way of phrasing 'if p then q' is as 'p only if q'. One way of thinking about the relationship between these two statements is to use the truth table shown in Table 7.1. The statement 'p only if q' is saying that p can be true only when q is, which is identical to saying that if p is true then q must be true as well.

Table 7.1   Truth table for the conditional
statements 'p only if q' and 'if p then q'.

p	q	p only if q	if p then q
T	T	T	T
T	F	F	F
F	T	T	T
F	F	T	T

Thinking about a contextual conditional may help convince you of the equivalence of these statements (although, as many researchers have demonstrated, adding context to conditionals can also lead reasoners astray, so care is needed). For instance, consider the conditional "if you are a policeman then you are tall". This is expressing a conditional relationship between being a policeman and being tall. It is quite reasonable to rephrase this as saying "you are a policeman only if you are tall". A similar example was used by Jonathan Evans to demonstrate that, although these statements are logically identical, they may not be cognitively identical.[1]

Evans hypothesised that the 'if...then' and 'only if' rules are interpreted differently by most people. His argument was based on the principles of necessity and sufficiency. Under the material understanding of 'if p then q', the p is sufficient for the q and the q is necessary for the p. However, these two principles are differentially emphasised in 'if...then' and 'only if' statements. Let's consider Evans's version of the policeman's height conditional. The rule 'If he is a policeman then he is over 5ft 9in in height' seems to emphasise the sufficiency of p, the statement 'he is a policeman'. On the other hand, the equivalent 'only if' rule, 'He is a policeman only if he is over 5ft 9in in height', seems to emphasise the necessity of q, the claim about height.[2] Based on this difference, Evans hypothesised two things. First, that more MP inferences would be made on 'if...then' statements than 'only if' statements, because the minor premise in MP affirms p, whose sufficiency is emphasised in 'if...then' statements. And second, that more MT inferences would be made on 'only if' statements than 'if...then' statements, because the minor premise in MT negates q, whose necessity is emphasised in 'only if' statements (when the necessary q is negated it is more obvious from 'only if' statements that p must also be negated).

---

[1] Evans, 1977.
[2] Evans, 1977, p. 300.

Table 7.2   Percentage endorsement rates for each
'if...then' and 'only if' inference in Evans's study.

	endorsed (%)	
Inference	'if...then'	'only if'
MP	100	76
DA	38	38
AC	67	84
MT	42	59

Evans investigated these ideas in a study with undergraduates (not from mathematics courses). His predictions were born out. MP inferences were endorsed 100% of the time in the 'if...then' condition, compared to 76% in the 'only if' condition. MT inferences, on the other hand, were endorsed 42% of the time in the 'if...then' condition compared to 59% of the time in the 'only if' condition.

This equivalency between 'only if' and 'if...then' statements gives a route with which we can test the extent to which the mathematics students' development identified in previous chapters generalises within the domain of logic. Do mathematics students — who we already know are more defective and less biconditional when reasoning with 'if...then' statements — show the same pattern on 'only if' statements? This question is important, because of the relative infrequency of 'only if' conditionals in mathematical writing.

7.1   'Only If' in Mathematical English

The logical equivalence of 'if p then q' and 'p only if q' is particularly helpful for addressing whether the primary mechanism for the TFD is exposure to relevant logical statements, or something deeper. This is because conditionals in mathematics are almost never written using the 'p only if q' form. Although this fact will be clear to anyone who has experience of formal written mathematics, it can also be demonstrated empirically.

Understanding the frequency with which various phrases occur in language, and what this means, is one of the goals of a discipline called corpus linguistics. Linguists working in this area construct extremely large collections of text, known as corpora, which they analyse using various statistical methods. For instance, suppose one wanted to determine the frequency of the two-word term 'only if' in general English. One could construct a large corpus of representative English texts and simply count the occurrences. In practice, of course, constructing corpora is extremely hard work: you have to be careful to select a balance of genres, a balance of written and spoken texts, and so on. Happily, several large representative corpora of general English exist.

For instance, searching for 'only if' in the 520 million words of the Corpus of Contemporary American English (COCA), reveals 5806 hits. So you would expect to see around 0.011 occurrences of 'only if' in every 1000 words of general American English. The frequency in British English, as revealed by a search of the British National Corpus, is slightly higher, at 0.015 occurrences per 1000 words.

What about in mathematical English? To investigate this question we, in collaboration with Lara Alcock, Kristen Lew, Pablo Mejía-Ramos, Paolo Rago and Chris Sangwin, decided to construct two corpora of mathematical texts. The first was based on mathematics research papers, the traditional mechanism through which mathematicians communicate their research findings to their colleagues. Conveniently for us, mathematicians have a culture of openly sharing pre-prints. These are drafts of formal papers that are made available before they have been officially published by academic journals. Typically nowadays, mathematicians upload their preprints to the ArXiv (pronounced 'archive'), a website designed to host academic texts.

Mathematical texts are normally produced, like this book, using a typesetting system called LATEX, which is similar in form to markup languages such as html. The source files for mathematics papers therefore contain both the paper's text and the various commands that the author used to produce mathematical symbols and equations. We downloaded the raw source of all 6988 mathematics papers posted to the ArXiv in 2009, and processed them to remove all LATEX commands and mathematical content. We were left with a corpus of 27 million words of formal mathematical English.

We also created a second corpus, using the conversations posted on the mathoverflow website. Mathoverflow is a site where mathematicians go to post and answer research-level mathematics questions. The site is very popular: in it's first three years it accumulated 16,000 users and hosted 27,000 conversations.[3] Ursula Martin and Alison Pease studied the types of interactions occurring on the site, concluding that "a typical interaction on mathoverflow is an informal dialogue, rather than rigorous steps of a correct

---

[3] Martin & Pease, 2013.

mathematical inference."[4] They found that the majority of questions, 64%, asked for factual information, such as references to proofs or theorems. But a substantial minority, 34%, were open-ended questions that sought to understand a piece of mathematics or to probe its motivation. So, compared to our ArXiv corpus, the text on mathoverflow can be thought of as a less formal version of mathematical English. We downloaded all mathoverflow posts made to date, and again stripped them of their mathematical symbols. This left us with 18 million words of informal mathematical English.

We were interested in understanding how the phrase 'only if' is used in mathematical communication. In particular, is it used to convey a meaning equivalent to 'if p then q', or is it only used for other purposes? To test this, we counted all occurrences of 'only if' and also 'if and only if'. Recall that 'p if and only if q' is a biconditional, equivalent to saying 'if p then q, and if q then p'. If the majority of occurrences of 'only if' in mathematics are as part of the phrase 'if and only if', then it seems reasonable to say that mathematicians very rarely use 'only if' to represent the material conditional 'if p then q'.

Our findings are shown in Figure 7.1. Over 90% of cases of 'only if' in both formal and informal mathematical English occur as part of the phrase 'if and only if'. In contrast, the reverse pattern is seen in general English: very few occurrences of 'only if' come from the phrase 'if and only if'.

The conclusion to take from this is that mathematics students are extremely unlikely to be exposed to conditionals expressed using 'only if' statements during their mathematical studies. If they see 'only if' it seems very likely to be part of the biconditional 'if and only if'. How then do mathematicians reason about 'only if' conditionals?

7.2 Reasoning with 'If ... Then' and 'Only If'

We recruited 61 third year mathematics undergraduates at a well-respected UK university to take part in a short study. The group was randomly split into two: half were assigned to be in the 'if...then' condition, the other half were assigned to be in the 'only if' group. The 'if...then' group were asked to solve the 32-item version of Evans's abstract conditional inference task, of the type used in the studies discussed earlier in the book. The 'only if' group had an identical task, except that their problems were phrased using

---

[4]Martin & Pease, 2013, p. 1.

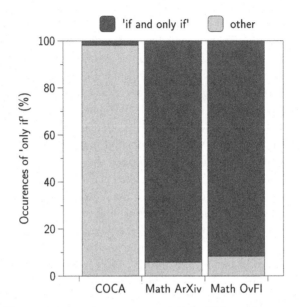

Figure 7.1 The percentage of occurrences of 'only if' in various corpora that are (i) part of 'if and only if' and (ii) part of some other sentence. COCA is the Corpus of Contemporary American English, Math ArXiv is our corpus of all 2009 mathematics preprints posted to the ArXiv in 2009, and Math OvFl is our corpus of discussions from mathoverflow.

the 'only if' form. For example, the 'if...then' group may have been asked to assess the validity of this inference:

> If the letter is B then the number is 6.
> The number is not 6.
> Therefore, the letter is not B.

Whereas the 'only if' group may have been asked about this inference:

> The letter is B only if the number is 6.
> The number is not 6.
> Therefore, the letter is not B.

The average number of inferences of each type endorsed by each group is shown in Figure 7.2. It is clear that there were substantial differences between the groups.[5] Compared to the 'if...then' group, those students

---

[5]A 2×4 ANOVA revealed a significant condition by inference-type interaction, $F(3, 177) = 16.07, p < 0.001, \eta_p^2 = 0.21$.

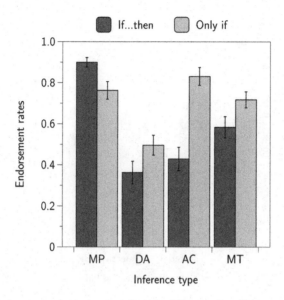

Figure 7.2 Endorsement rates for the MP, DA, AC and MT inferences for each group. Error bars show ±1 SE of the mean.

who were asked to solve 'only if' problems endorsed fewer MP inferences,[6] more AC[7] and MT[8] inferences, and slightly more DA inferences.[9]

Perhaps a clearer way of understanding the differences between the two groups is to assess which interpretation of the conditional their responses were most consistent with. The extent to which their responses were consistent with the material, defective and biconditional interpretations is shown in Figure 7.3.

In earlier chapters, we saw that a feature of mathematics students' conditional reasoning was a reduction in the frequency of biconditional responses, and an increase in the frequency of material and, particularly, defective responses. Figure 7.3 paints a clear picture: the mathematics students in the 'only if' group had responses more consistent with how non-mathematics students reason with 'if...then' statements. The pattern of responses we have come to associate with mathematicians was only present here with those who reasoned with conditionals phrased in the manner normally used in mathematical texts: 'if p then q'. The implication

---

[6] $t(59) = 2.82, p = 0.007, d = 0.72.$
[7] $t(59) = 5.58, p < 0.001, d = -1.43.$
[8] $t(59) = 2.09, p = 0.041, d = -0.53.$
[9] $t(59) = 1.81, p = 0.075, d = -0.46.$

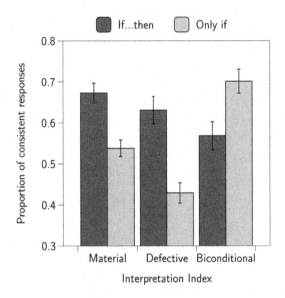

Figure 7.3   The extent to which the two groups responded in line with the material, defective and biconditional interpretations. Error bars show ±1 SE of the mean.

seems clear. The advantage shown by mathematics students on conditional reasoning is only present, or at least only present to such a large degree, when they are asked to reason with conditionals expressed in the manner most commonly used in mathematical texts.

At this point a sceptical reader might be wondering about our earlier remark about the lack of explicit 'if p then q' statements in A level mathematics exams. Recall that we examined every first year A level mathematics examination set between 2009 and 2011, and found that of 929 questions set, only one contained an explicit "if...then" sentence. In this chapter, we have argued that a plausible mechanism behind the reasoning development shown by mathematics students is their regular exposure to 'if...then' statements, but if our A level examination analysis is to be believed, this might not be the case.

Our suggested explanation for this apparent contradiction is that although A level mathematics students may not see explicit 'if...then' statements that regularly, they *do* regularly experience forward deductions where p being true is sufficient for q to be true. Indeed, any algebraic manipulation takes essentially this form. Consider the following question, taken from a recent A level mathematics examination paper:

Solve the inequality $6x^2 \leq x + 12$.

How might a student solve this? A typical response might go something like:

We know that $6x^2 \leq x + 12$

$$\implies 6x^2 + x - 12 \leq 0$$
$$\implies (3x + 4)(2x - 3) \leq 0$$
$$\implies \text{The critical values are } x = -\frac{4}{3}, \frac{3}{2}.$$

When $x < -\frac{4}{3}$ and $x > \frac{3}{2}$, we have $6x^2 + x - 12 > 0$

So the solution is $-\frac{4}{3} \leq x \leq \frac{3}{2}$.

Notice that at each stage in this solution, the truth of line n is sufficient to conclude the truth of line $n + 1$. So although no explicit 'if...then' statement is included in this solution, a large number of implicit 'if...then' statements are included. In contrast, none of these deductions emphasise that the truth of line $n + 1$ is necessary for the truth of line n: none of these deductions are implicit 'only if' conditionals. This type of algebraic manipulation seems to be extremely common in A level mathematics, and so students studying the course regularly read and write statements that, like 'if p then q', emphasise the sufficiency of p for the truth of q. We suspect that it is this repeated exposure to forward deductions that is behind the reasoning development observed in post-compulsory mathematics students reported in earlier chapters.

It is worth noting that a failure to accurately distinguish between the conditional and the biconditional can be the source of many mistaken lines of reasoning in post-compulsory mathematics. Perhaps the easiest way of demonstrating this is to give an example of a fallacious proof.[10] Here, for instance, is a 'proof' that demonstrates that $3 = 5$:

We know that $9 - 24 = 25 - 40$.

Therefore $9 - 24 + 16 = 25 - 40 + 16$.

Factoring both sides gives $(3 - 4)^2 = (5 - 4)^2$.

$$\implies 3 - 4 = 5 - 4.$$

Therefore, by adding 4 to both sides, we know that $3 = 5$.

---

[10]We are grateful to Chris Sangwin for sharing these examples, from Maxwell (1959), with us.

At first glance this argument looks reasonable. But, clearly, 3 does not equal 5, so what has gone wrong? The answer is that the argument relies upon someone mixing up a conditional and biconditional. Line 4 implicitly relies upon the conditional 'if $x^2 = y^2$ then $x = y$', which is false in general: taking $x = -2$ and $y = 2$ provides a counterexample ($(-2)^2 = 2^2$, but $-2 \neq 2$). While mathematics students will be familiar with the true conditional 'if $x = y$ then $x^2 = y^2$', if they inappropriately interpret this as being the biconditional '$x = y$ if and only if $x^2 = y^2$', then this may lead to the kind of logical paradoxes that underlie this argument.

Here is another example, which relies upon material taught in A level mathematics[11]:

Let $I = \int \frac{1}{x} \, \mathrm{d}x$, and we will integrate by parts.

$$I = \int 1 \times \frac{1}{x} \mathrm{d}x$$
$$= x\frac{1}{x} - \int x\frac{-1}{x^2}\mathrm{d}x$$
$$= 1 + \int \frac{1}{x}\mathrm{d}x$$
$$= 1 + I.$$

So, if we subtract I from both sides, we have $0 = 1$.

The interested reader is encouraged to spot why this argument relies upon mixing up 'if p then q' with 'p if and only if q'.

Our proposal here is in fact extremely similar to that put forward by early proponents of the TFD. In Chapter 1 we cited J. G. Fitch's support for the TFD as expressed in his 1880 'Lectures on Teaching', delivered at the University of Cambridge. Describing mathematics, he wrote:

> Here, at least, the student moves from step to step, from premiss to inference, from the known to the hitherto unknown, from antecedent to consequent, with a firm and assured tread; knowing well that he is in the presence of the highest certitude of which the human intelligence is capable.[12]

It is this repeated forwards step-to-step journey from premise to inference, or antecedent to consequent, that seems to be so common in advanced

---

[11] Also from Maxwell (1959), via Chris Sangwin.
[12] Fitch, 1883, p. 343.

mathematics. Given this, it is perhaps unsurprising to find that the reasoning gains associated with mathematical study are primarily associated with those conditionals that emphasise that p is sufficient to conclude q ('if p then q'), rather than back-to-front conditionals that emphasise that q is necessary for p ('p only if q').

Chapter 8

Summary and Conclusions

8.1 Evidence Presented in the Book

Our goal in this book has been to investigate the TFD, the idea that studying advanced mathematics develops students' "general thinking skills" that are useful in many areas of life. Our investigation was motivated by the strange situation outlined in Chapter 1. There we suggested that the TFD was endorsed by a great many influential philosophers, mathematicians, and educational policymakers. But, alongside this confident endorsement of the TFD, we pointed out that the majority view among psychologists, established from over 100 years of research on transfer in education, was that the TFD is simply false.

We provided four sources of evidence for our claim that the TFD is widely endorsed by the mathematical community. First, we quoted a large number of important historical thinkers who explicitly endorsed the TFD. Perhaps the clearest exposition of the theory came from Plato, who wrote that

> Those who have a natural talent for calculation are generally quick at every other kind of knowledge; and even the dull, if they have had an arithmetical training, although they may derive no other advantage from it, always become much quicker than they would otherwise have been.[1]

Alongside Plato, we saw that similar comments have been made by John Locke, Isaac Watts, Francis Bacon, and many others.

Second, we reported that the TFD had been cited in a great number of reports that were designed to influence government policy on mathematics education. These include the Cockcroft Report from the early 1980s, the Smith Report from the mid 2000s, and the Vorderman Report from 2010. Each, to a greater or lesser extent, justified the privileged place that

---

[1] Plato, 375BC/2003, p. 256.

mathematics has in the school curriculum by appealing to its ability to develop students' reasoning skills.

Third, we conducted a UK-wide survey of recent mathematics graduates to determine their views on the value of mathematical study. We found that the vast majority — well over 90% — felt that their mathematics degree had developed their ability to think logically and analytically, and also their problem-solving skills.

Finally, we interviewed a number of influential policymakers about their views on the place of mathematics in the school curriculum. All enthusiastically endorsed the proposal at the heart of the TFD. Although alongside this enthusiasm there was some recognition that there might be an alternative explanation for the observation that mathematics students are 'better' at reasoning than their non-mathematical friends.

We then reviewed some of the extensive research evidence on educational transfer. The issue is this: can skills or concepts learned in one context be successfully applied in another? We saw that, since the work of Edward Thorndike in the early twentieth century, psychologists have been highly sceptical of the notion of transfer in general, and of transfer to logic in particular. Notably, Patricia Cheng and Richard Nisbett in the 1980s and 1990s found that even studying formal logic didn't seem to improve students' chances of correctly solving logical problems (or at least, Wason selection tasks).

This apparent conflict between the weight of opinion in the mathematical community, and the weight of opinion among research psychologists was the starting point of our investigation. Our research strategy followed a four-step process. First, we considered how best to measure the reasoning gains that proponents of the TFD hypothesise would occur following mathematical study. We interviewed stakeholders in the mathematical community to get their views about the specific reasoning tasks that they felt mathematical study would facilitate. This process identified three specific tasks: the conditional inference task, the Aristotelian syllogism task, and the Wason selection task.

Second, we conducted a cross-sectional study to compare the reasoning behaviour of mathematics undergraduates and English literature undergraduates. If the TFD were correct, we would have expected to have seen substantial between-group differences with the mathematics group having an advantage on all tasks. This pattern of results was essentially what we found. In the first of the two cross-sectional studies reported in Chapter 3, we found that a between-groups difference was present across all of our tasks, but that it was larger on problems phrased using abstract content, and that it was larger on the conditional inference task than on the Aristotelian syllogism task. In a second study, we asked the same question about the Wason selection task, and again found a between-group difference in reasoning behaviour, although in this case the result was rather more nuanced.

The TFD could have been refuted with the data from our cross-sectional studies. If we had found no difference in the reasoning behaviour of the mathematics and literature groups, then this would have caused great difficulty for the theory. However, the fact that we did find such a difference is consistent with both the TFD and its main rival, the filtering hypothesis. Perhaps those students who choose to study post-compulsory mathematics already reason differently to those who do not?

Given this, the third stage of our research strategy was to conduct a longitudinal study designed to try to separate the TFD from the filtering hypothesis. In the first study reported in Chapter 4, we tracked groups of A level mathematics and English literature students over their first year of post-compulsory study. We focused on abstract conditional inference, since this was where the largest between-groups difference in the cross-sectional study was found. Critically, we found that our two groups didn't differ in conditional inference behaviour at the start of their A level studies. This observation is extremely important as it means that the Matthew effect is unlikely to be operating here. If we had found an existing group difference at time 1, and if this difference had widened over the course of a year of study, then this might simply have been a case of 'the rich getting richer', rather than anything related to mathematical study. The fact that we didn't find any pre-existing differences at the start of A level — but that by the end of the first year group differences had emerged — is highly consistent with the TFD, and difficult (albeit not impossible) to explain through the filtering hypothesis.

But the longitudinal study in Chapter 4 did throw up one puzzle. Apparently, the A level mathematics students actually became *less* normative on the modus tollens inference over time, in the sense that they were less likely to endorse MT inferences after studying A level mathematics for a year. From an actor-oriented transfer perspective, this doesn't really matter: we have good reason to suspect that this change in behaviour occurred due to the students' mathematical study, whether or not the change is desirable or undesirable is not so important from an actor-oriented perspective. But from the perspective of the TFD this result is rather strange: the mathematical stakeholders we interviewed in Chapter 1 were clear that they felt that mathematics students would reason more in line with the norms of formal logic, which in this case would mean developing a material interpretation of the conditional, not a defective interpretation. Or, to put it another way, TFD proponents expected mathematics students would be endorsing *more* MT inferences, not less. Because we were somewhat surprised by this result, we followed it up with a cross-sectional comparison of first year university students' behaviour on the truth table task. Again we found behaviour consistent with a defective interpretation of the conditional.

In the second longitudinal study reported in Chapter 4, we repeated our study in Cyprus. We again found no difference between our two groups' conditional reasoning behaviour at age 16, but over time differences did emerge. The results were consistent with the UK-based longitudinal study in the sense that the changes in behaviour in the high-intensity mathematics group came about through an increased rejection of the invalid DA and AC inferences. However, unlike in our English sample, we found that the high-intensity mathematics group didn't actually decline in MT endorsement rates, but neither did they increase.

Both of these longitudinal studies seem to favour the TFD over the filtering hypothesis. But we do not wish to be interpreted as concluding that the filtering hypothesis can be entirely dismissed. Throughout these studies we took care to check for the existence of a Matthew effect on cognitive capacity and thinking disposition. We checked whether or not our measures of Raven's matrices and the cognitive reflection test (CRT) predicted gains in conditional inference scores between the time points. Because they did not, we felt confident to rule out a Matthew effect operating via the mechanism of cognitive capacity or thinking disposition. However, these measures can also be conceptualised as outcome measures rather than covariates. In other words, we can check to see whether or not our mathematics and literature groups started post-compulsory studies with similar levels of performance on the Ravens and CRT measures, and whether they diverge during their studies. The data from the A level longitudinal study, reported in Chapter 4, are shown in Figure 8.1.

These graphs show, in contrast to the results from the conditional inference task, that our two groups entered post-compulsory study with pre-existing differences. Although both plots show a trend for the mathematics group to increase their scores at a slightly greater rate than the comparison group, neither of these effects reach statistical significance, suggesting that

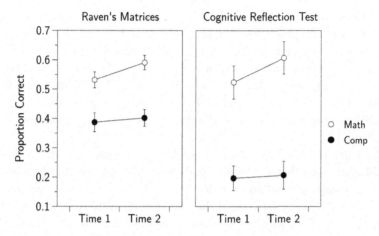

Figure 8.1 The proportion of Raven's and CRT items answered correctly by the mathematics and comparison (literature) groups in the A level longitudinal study at the start and end of the year. Error bars show ±1 SE of the mean.

we cannot conclude that the mathematics group improved to a greater extent than the comparison group.[2] Similar results were observed in the Cypriot study.[3]

So, although we have suggested that the TFD offers a better account of our main focus — related to conditional inference — than the filtering hypothesis, this should not be interpreted to mean that the filtering hypothesis has no merit at all. Indeed, it seems quite plausible to suppose that there are some general reasoning skills for which the TFD holds true, and there are some others for which the filtering hypothesis offers the better account.

In Chapter 5, we explored the surprising result that mathematics students apparently become less likely, or at least no more likely, to endorse the MT inference as a result of their mathematical studies. We found a similar pattern of development among mathematics undergraduates: whereas studying undergraduate mathematics is, like studying A level mathematics, apparently associated with an increased tendency to reject

---

[2]Raven's: time by group interaction, $F(1, 80) = 1.531, p = 0.220$. CRT: time by group interaction, $F(1, 79) = 2.013, p = 0.160$.

[3]Raven's: time by group interaction, $F < 1$. CRT: time by group interaction, $F(2, 360) = 3.188, p = 0.043$. Note that although this latter p does meet the conventional significance threshold of $\alpha = 0.05$, it would not if an appropriate Bonferroni correction were applied to control the familywise error rate.

invalid inferences, it apparently does not substantially change behaviour on MT problems. We also demonstrated that the extent to which students rejected the invalid DA and AC inferences was predictive of their performance on university mathematics courses, and on a specially designed measure of proof comprehension. In contrast, no such association was found for MT. Given this, we concluded that it appears to be feasible to succeed in university mathematics by adopting a defective interpretation of the conditional.

In Chapter 7, we asked why studying mathematics might develop conditional reasoning skills. By analysing the language mathematicians use, and comparing undergraduates' behaviour on 'if...then' and 'only if' inference problems, we suggested that mathematical study requires students to regularly follow forward inferences from premise to conclusion. The conclusion we reached was essentially identical to the position advanced by J. G. Fitch in his Cambridge lectures of 1880, who argued that in mathematics "the student moves from step to step, from premiss to inference, from the known to the hitherto unknown, from antecedent to consequent, with a firm and assured tread".[4]

8.2   Does Studying Mathematics Actually *Improve* Logical Thinking?

As we have repeated throughout the book, definitively concluding a causal relationship between post-compulsory mathematical study and reasoning development is extremely difficult. But, suppose we are willing to accept that this is the most plausible way of accounting for our data. Can we conclude that studying mathematics actually *improves* conditional reasoning? Answering this question requires us to reconsider the notion of norms. What constitutes 'good' reasoning?

Earlier in the book we noted that Stanovich argued that there are three broad perspectives you can take on reasoning development.[5] Which you adopt depends on your views on how people actually do reason (descriptive reasoning), how you think they should reason (normative reasoning), and the best level of reasoning that is possible, given our cognitive constraints (prescriptive reasoning). Panglossians assert that descriptive, prescriptive and normative reasoning are one and the same: humans already reason as

---

[4]Fitch, 1883, p. 343.
[5]Stanovich, 1999.

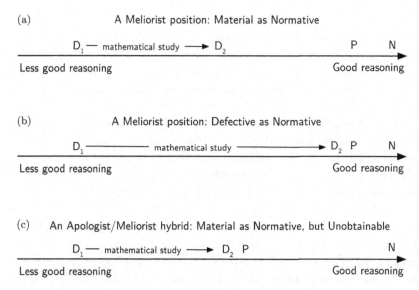

Figure 8.2 Three different proposals for the relationship between normative, descriptive and prescriptive reasoning, given the findings reported in this book. D_i shows the descriptive situation at time i, P shows the prescriptive position, and N shows the normative position.

well as they can, and this is also how they should reason. Adopting such a position is clearly inconsistent with the TFD, which explicitly endorses the idea that reasoning can be improved. Meliorists assert that human reasoning typically falls well short of what is both possible and desirable. Finally, Apologists assert that although it would be desirable for humans to reason normatively, it is beyond our ability to do so. These three positions were summarised in Figure 2.2.

What positions are now available to us, given the findings reported in Chapters 3–7? Figure 8.2 illustrates three possible stances. If we retained the material interpretation of the conditional as the normative and prescriptive standard, as the stakeholders interviewed in Chapter 2 appeared to do, then mathematical study is only partially effective. Although it does appear to increase the likelihood of invalid inferences being rejected, it may also decrease (or at least not increase) the likelihood of the valid MT inference being endorsed. In other words, mathematical study moves students towards the prescriptive and normative end of the reasoning spectrum, but they are still some way away. This perspective is shown in Figure 8.2(a).

In contrast, if we accepted the arguments made in Chapter 5 and concluded that the defective interpretation of the conditional can reasonably be said to be in some sense normative, then we would end up with the much more optimistic picture shown in Figure 8.2(b). In this case, mathematical study appears to be reasonably effective at developing reasoning skills as it is associated with declines in the endorsement of invalid inferences.

A third possibility is to retain the material conditional as the normative standard, but to assert that it is simply unobtainable. Perhaps it is unreasonable to expect the majority of reasoners, regardless of their mathematical background, to reliably endorse MT. In this case, one might adopt an Apologist/Meliorist hybrid position shown in Figure 8.2(c).

All three of these positions are consistent with the TFD in that they all suggest that studying mathematics improves reasoning skills. The difference between them concerns the extent of that improvement. We tend towards the optimistic Meliorist position, outlined in Figure 8.2(b), but the appropriateness of norms is not a matter that can be definitively settled by empirical investigation.

8.3   The 'Real World' Objection

In Chapter 2, we set out the 'incorrect norm' argument. Proponents of this position suggest that researchers who use laboratory reasoning tasks have systematically been evaluating their participants' behaviour using incorrect norms.[6] Some have taken this line of thought further and have argued that laboratory reasoning tasks, of the type that we have used throughout this book, have no obvious connection to real-world reasoning at all, and therefore that they should be avoided.[7] A person who adopted this perspective would presumably feel that, although the TFD might be correct to assert that studying mathematics develops certain general reasoning skills, it is incorrect to assert that these skills are useful in day-to-day life.

We don't find this critique of the psychology of reasoning literature persuasive. It can be criticised on both theoretical and empirical grounds. Keith Stanovich summarised the theoretical objection when he wrote:

> The issue is that, ironically, the argument that the laboratory tasks and tests are not like 'real life' is becoming less and less true. 'Life,' in fact, is becoming more like the tests![8]

---

[6]Cohen, 1981; Lopes, 1991.
[7]For example, Reid in Reid and Inglis (2005).
[8]Stanovich, 2004, p. 124.

What Stanovich meant is that while decontextualised reasoning may have been unimportant in the environment in which the human brain evolved — the African savanna — the same cannot be said for a modern post-industrial society. If you live in a band of hunter gatherers it might be appropriate to reason based on your Type 1 emotional reactions or your Type 1 intuitions of fairness. But this approach simply will not work if you need to determine whether you are eligible for a tax rebate, or whether or not your insurance policy covers your partner when driving a car abroad. These are situations governed by abstract rules, and successfully negotiating them requires the ability to decontextualise, ignore Type 1 responses, and carefully use Type 2 reasoning to explore the consequences of the rules.

Rules that govern even mundane day-to-day situations are often written using breathtakingly complex logical structures. Consider, for instance, Law 32 in cricket, which governs catches. One might think that this is a relatively simple matter: if a fielder catches the ball after it has been struck by the batsman, and before it touches the ground, then the batsman is out. However, things are not so straightforward. Subsection 3 of the Law 32 is actually phrased like this:

> Providing that in every case neither (i) at any time the ball, nor (ii) throughout the act of making the catch as defined in Law 19.4, any fielder in contact with the ball is, as described in Law 19.3(b), touching the boundary or grounded beyond the boundary, a catch shall be considered to be fair if
>
> (a) the ball is hugged to the body of the catcher or accidentally lodges in his clothing or, in the case of a wicket-keeper only, in his pads. However, it is not a fair catch if the ball lodges in a protective helmet worn by a fielder.
> (b) the ball does not touch the ground even though the hand holding it does so in effecting the catch.
> (c) a fielder catches the ball after it has been lawfully struck more than once by the striker, but only if it has not been grounded since it was first struck.
> (d) a fielder catches the ball after it has touched an umpire, another fielder or the other batsman. However, it is not a fair catch if at any time after having been struck by the bat and before a catch is completed the ball has touched a protective helmet worn by a fielder.
> (e) a fielder catches the ball after it has crossed the boundary in the air, provided that after being struck by the bat, the first contact with the ball is by a fielder, not touching or grounded beyond the boundary, who has some part of his person grounded within the boundary or whose final contact with the ground before touching

the ball was entirely within the boundary. Any fielder subsequently touching the ball is not subject to this restriction. See Law 19.4 (Ball beyond the boundary).

(f) the ball is caught off an obstruction within the boundary that has not been designated a boundary by the umpires before the toss.

This is only one subsection out of five, and Law 32 is only one law out of the 42 that govern cricket! Perhaps an amateur cricket player might be able to get away without understanding the details of the laws of the game — although they might live in a state of perpetual confusion — but it seems clear that any aspiring umpire is not likely to be successful if they cannot successfully deal with complex conditional statements.

Many similar examples can easily be found. For instance, would a homeowner in England require the permission of their local council to build a new shed in their garden? The answer is given by Section A.1 of Part 1 of Schedule 2 of the Town and Country Planning (General Permitted Development) (England) Order 2015. The section contains four pages of dense abstract rules including conditional statements, disjunctive statements, subclauses, and references to technical terms (e.g. "dwellinghouse", "curtilage") that each have precise definitions.[9]

Fluency with the interpretation of, and indeed the construction of, complex logical rules is, argued Stanovich, an increasingly important skill in a knowledge-based economy. He summed up the situation by writing:

> Given the ubiquitousness of such abstract directives in our information and technology-saturated society, it just seems perverse to argue the "unnaturalness" of decontextualised reasoning skills when such skills are absolutely necessary to success in our society.[10]

He went further by pointing out the irony of some academic critics developing arguments against the importance of Type 2 abstract reasoning skills:

> For intellectuals to use their abstract reasoning skills to argue that the "person in the street" is in no need of such skills of abstraction is like a rich person telling someone in poverty that money is not really all that important.[11]

If Stanovich is right about this, then we ought to be able to find evidence that fluency with decontextualised reasoning problems is in some sense

---

[9]http://www.legislation.gov.uk/uksi/2015/596/contents/made.
[10]Stanovich, 2004, p. 125.
[11]Stanovich, 2004, p. 125.

Part 1. Imagine that recent evidence has shown that a pesticide is threatening the lives of 1200 endangered animals. Two response options have been suggested:

(1) If Option A is used, 600 animals will be saved for sure.
(2) If Option B is used, there is a 75% chance that 800 animals will be saved and a 25% chance that no animals will be saved.

Which option do you recommend to use?

Part 2. Imagine that recent evidence has shown that a pesticide is threatening the lives of 1200 endangered animals. Two response options have been suggested:

(1) If Option A is used, 600 animals will be lost for sure.
(2) If Option B is used, there is a 75% chance that 400 animals will be lost and a 25% chance that 1200 animals will be lost.

Which option do you recommend to use?

Figure 8.3   Two parts of the 'Resistance to Framing' problem.

predictive of success in our society. In fact such evidence does exist, and it provides the second reason to doubt the legitimacy of those who seek to dispute the relevance of laboratory reasoning tasks.

Perhaps the most persuasive demonstration of the relevance of abstract reasoning skills to day-to-day life was presented by Wändi Bruine de Bruin and colleagues in 2007.[12] They created a battery of laboratory reasoning tasks, of the type criticised by proponents of the 'incorrect norm' argument. For example, one problem they used is shown in Figure 8.3.[13]

The two problems shown in Figure 8.3 are logically identical, but if you give both problems to the same person (with different problems in between, to reduce the chances of them noticing the similarity) it is common to receive different responses. People typically avoid risk when a positive outcome is available (600 animals will be saved) and seek risk when a negative outcome is likely (600 animals will be lost). This effect is normally accounted for by Amos Tversky and Daniel Kahneman's Prospect Theory,

---

[12]Bruine de Bruin, Parker, & Fischoff, 2007.
[13]Adapted from Tversky and Kahneman (1981).

which states that losses are psychologically more significant than their equivalent gains. Bruine de Bruin and colleagues gave participants a series of problems like this, which ended up with them receiving a 'Decision Making Competence' score, which assessed the extent to which they responded normatively. In the example given in Figure 8.3, the maximum possible score would be obtained by a participant who gave identical responses to parts 1 and 2.

But to assess whether or not performance on these kinds of abstract problems is related to real world outcomes, the researchers also needed some kind of measure of how successful an individuals' day-to-day decisions are. Their approach was to develop the ingenious 'Decision Outcomes Inventory' (DOI).[14] The DOI attempts to measure the extent to which individuals have made obviously bad decisions over the course of their lives, using self reports. For instance, if you have ever gone shopping for food or groceries (99.4% of participants in Bruine de Bruin's study had), then it is clearly undesirable for you to have needed to throw your purchases out because they went bad (81% of participants in the study had done this at least once). Similarly, it is clearly undesirable to miss a flight (a bad outcome that 7% of participants had suffered), take the wrong train (8.3%), have a cheque bounce (32%), or drink so much alcohol that you vomited (28%).

All of these negative outcomes can be attributed, to some degree at least, to poor 'real life' decision making skills. So the DOI provides a quantitative measure of how often an individual suffers bad outcomes due to poor decision making. It also allows us to estimate prevalence rates of various different errors of judgement. For instance, 37% of people in Bruine de Bruin's study had apparently been locked out of their house, 7% had been kept overnight in a jail cell, and 22% had suffered blisters as a result of sunburn.

Using responses to these kinds of items, Bruine de Bruin calculated each participant's DOI score, a measure of the extent to which their decision making led to bad real-world outcomes. The key finding from the study was that individuals' DOI scores correlated positively with their overall score on the battery of decision-making tasks.[15] Importantly, this correlation did not change a great deal when the researchers controlled for participants' cognitive capacity, as measured by Raven's matrices and a

---

[14]Bruine de Bruin *et al.*, 2007.
[15]$r = 0.29, p < 0.001$.

reading comprehension test.[16] In other words, differences in performance on the laboratory decision-making tasks predicted real-world decision making over and above general cognitive capacity.

This finding can also be broken down by the different real-world outcomes. When the researchers split their sample into two — those who scored well on the laboratory decision-making tasks and those who scored poorly — they found substantial differences in the frequency of poor outcomes.[17] For instance, 35% of the low-scoring group had once needed to replace the key to their home because they had lost it, compared to only 13% of the high-scoring group. Similarly, 23% of the low-scoring group had quit a job after only a week, compared to just 3% of the high-scoring group.

These results are extremely problematic for those who wish to deny the relevance of 'artificial' laboratory reasoning tasks to real world behaviour. If such tasks did not index any kind of skill that is pertinent to day-to-day life, then it would be extremely difficult to account for these findings. Perhaps one reasonable way of objecting to the relevance of this work might be to suggest that the kinds of laboratory decision-making tasks used by Bruine de Bruin and colleagues are different to the reasoning tasks we have concentrated on in this book. But there are counterarguments that can be deployed against such an critique. Perhaps most convincingly, Keith Stanovich and Richard West have shown that performance on laboratory conditional reasoning tasks tends to correlate with performance on laboratory decision-making tasks similar to the types used by Bruine de Bruin and colleagues.[18]

In summary, we are unconvinced by the suggestion that laboratory reasoning tasks are entirely unrelated to real world outcomes. Here we have offered two main reasons. First, successfully engaging with the kinds of rules that govern post-industrial societies does seem to require the ability to reason fluently with abstract rules. And second, there is empirical evidence which links performance on 'artificial' laboratory tasks with real-world outcomes.

8.4 Concluding Remarks

We begun our investigation of the issues discussed in this book genuinely uncertain about the validity of the TFD. On the one hand, the weight of

---

[16]$pr = 0.26, p < 0.001.$
[17]Parker, Bruine de Bruin, & Fischoff, 2015.
[18]Stanovich & West, 1998; Stanovich & West, 2000.

opinion in its favour seemed large and authoritative. On the other, research evidence from the educational psychology literature seemed to cast serious doubt upon it. If our starting point was uncertainty, how best can we characterise our view now that we have reached the end of this series of studies?

We believe that the evidence we have presented provides good reasons to suppose that studying advanced mathematics does indeed develop some aspects of conditional reasoning, notably the ability to reject invalid inferences. In that sense, we have found evidence that is highly consistent with at least one version of the TFD. This, coupled with evidence that the jobs market continues to value those with an advanced education in mathematics,[19] suggests that the TFD is in robust health. Perhaps Thorndike's critiques of the TFD were unduly pessimistic.

However, TFD supporters should perhaps not be entirely enthused by the evidence we have presented. Although we saw that the mathematics students in our studies appeared to develop their conditional reasoning, this was a more nuanced development than that predicted by traditional versions of the TFD. Specifically, mathematics students seemed to improve their detection of invalid inferences, not their overall conditional inference. We also found that this development did not generalise to all linguistic formulations of conditional statements. Finally, we found that there appear to be some cognitive skills — those required for a high-level of performance on Raven's matrices and the cognitive reflection test — for which the filtering hypothesis seems to provide a better account of the gap in performance between mathematicians and non-mathematicians.

Overall then, what is the answer to the question posed in our title? Does mathematical study develop logical thinking? We believe that it does, although this development is much more nuanced than often claimed by mathematicians and policymakers.

---

[19] Adkins & Noyes, 2016; Dolton & Vignoles, 2002.

Appendix A

Reasoning Tasks

Participants in the interview study (reported in Chapters 1 and 2) were given the following information about each task and asked (for each) to state the extent to which they agreed with the following statement: "this task captures some of the skills that studying advanced mathematics develops".

Argument Evaluation Task

Aim: To assess participants' ability to evaluate arguments independently of their prior belief.

Instructions:

We are interested in your ability to evaluate counter-arguments. First, you will be presented with a belief held by an individual named Dale. Following this, you will be presented with Dale's justification for holding this particular belief. A Critic will then offer a counter-argument to Dale's justification. Finally, Dale will offer a rebuttal to the Critic's counter-argument. You are to evaluate the strength of Dale's rebuttal, regardless of your feeling about the original belief or Dale's premise. You should assume all statements are factually correct.

Example Problem:

Dale's belief: It is more dangerous to travel by air than by car.

Dale's justification: Air accidents are more likely than car accidents to involve fatalities.

Critic's counter-argument: Passengers are three times more likely to be killed per mile travelled in a car as compared to a plane.

Dale's rebuttal: Because reckless or drunk drivers cause the great majority of all automobile accidents, car travel is at least safer than air travel for people who wear safety belts and travel with sober and careful drivers.

Indicate the strength of Dale's rebuttal to the Critic's counter-argument:

○ Very Weak   ○ Weak   ○ Strong   ○ Very Strong

Figure A.1   The argument evaluation task.

Source: Stanovich, K. E., & West, R. F. (1998). Individual differences in rational thought. *Journal of Experimental Psychology: General,* 127, 161–188.

Belief Bias Syllogism Task

Aim: To assess whether participants are able to evaluate logical syllogisms independently of their prior beliefs.

Instructions: In the following problems, you will be given two premises, which you must assume are true. A conclusion from the premises then follows. *You must decide whether the conclusion follows logically from the premises or not. You must suppose that the premises are all true and limit yourself only to the information contained in the premises.*

Some example problems:

- No addictive things are inexpensive.
  Some cigarettes are inexpensive.
  Therefore, some addictive things are not cigarettes.
  ○ YES   ○ NO

- No millionaires are hard workers.
  Some rich people are hard workers.
  Therefore, some millionaires are not rich people.
  ○ YES   ○ NO

- No police dogs are vicious.
  Some trained dogs are vicious.
  Therefore, some trained dogs are not police dogs.
  ○ YES   ○ NO

- No nutritional things are inexpensive.
  Some vitamin tablets are inexpensive.
  Therefore, some vitamin tablets are not nutritional.
  ○ YES   ○ NO

Figure A.2 The belief bias syllogism task.
Source: Evans, J., Barston, J., & Pollard, P. (1983). On the conflict between logic and belief in syllogistic reasoning. *Memory & Cognition,* 11, 295–306.

Cognitive Reflection Task

Aim: To assess whether participants are able to override intuitively appealing but incorrect answers to simple arithmetic problems.

Instructions: This section consists of three problems of varying difficulty. Answer as many as you can.

Problems:

- A bat and ball costs £1.10 in total. The bat costs £1.00 more than the ball. How much does the ball cost? ... .pence.

- If it takes 5 machines 5 minutes to make 5 widgets, how long would it take 100 machines to make 100 widgets? ....minutes.

- In a lake, there is a patch of lily pads. Every day, the patch doubles in size. If it takes 48 days for the patch to cover the entire lake, how long would it take for the patch to cover half of the lake? ....days.

Figure A.3   The cognitive reflection task.

Source: Frederick, S. (2005). Cognitive re ection and decision making. *Journal of Economic Perspectives*, 19, 25–42.

Conditional Inference Task

Aim: To assess whether participants are able to determine whether a conclusion follows from an (abstract) conditional statement and premise.

Instructions: Each problem concerns an imaginary letter-number pair and contains an initial statement or rule which determines which letters may be paired with which numbers. The task in each case is to decide whether or not the conclusion *necessarily* follows from the statements. A conclusion is necessary if it must be true, given that the statements are true.

Some example problems:

- If the letter is A then the number is 3.

  The letter is A.

  Conclusion: The number is 3.

  ○ YES   ○ NO

- If the letter is D then the number is not 4.

  The number is 4.

  Conclusion: The letter is not D.

  ○ YES   ○ NO

- If the letter is not H then the number is 1.

  The number is not 1.

  Conclusion: The letter is H.

  ○ YES   ○ NO

- If the letter is K then the number is not 3.

  The number is not 3.

  Conclusion: The letter is K.

  ○ YES   ○ NO

Figure A.4   The conditional inference task.

Source: Evans, J., & Handley, S. (1999). The role of negation in conditional inference. *Quarterly Journal of Experimental Psychology,* 52A, 739–769.

Evaluation of Arguments (Watson-Glaser Critical Thinking Appraisal)

Aim: To assess participants' ability to discriminate between strong and weak, important and irrelevant arguments.

Instructions: In making decisions about important questions, it is desirable to be able to distinguish between strong and weak arguments. For an argument to be strong it must be both important, and directly related to the question. Below is a series of questions. Each question is followed by several arguments. *You should regard each argument as true.* The problem is to decide whether it is a STRONG or a WEAK argument.

An example problem:

Should groups in this country who are opposed to some of our governments' policies be permitted unrestricted freedom of press and speech?

- Yes; a democratic state thrives on free and unrestricted discussion, including criticism.

  ○ STRONG   ○ WEAK

- No; the countries opposed to our form of government do not permit the free expression of our point of view in their territory.

  ○ STRONG   ○ WEAK

- No; if given full freedom of press and speech, opposition groups would cause serious internal strife and make our government basically unstable, eventually leading to the loss of our democracy.

  ○ STRONG   ○ WEAK

Watson, G. & Glaser, E. M. (1964). *Watson-Glaser Critical Thinking Appraisal.* Hardcourt, Brace & World

Figure A.5   Evaluation of arguments (Watson–Glaser critical thinking appraisal).
Source: Watson, G. & Glaser, E. M. (1964). *Watson–Glaser Critical Thinking Appraisal.* San Diego: Hardcourt, Brace & World.

Interpretation (Watson–Glaser Critical Thinking Appraisal)

Aim: To assess participants' ability to weigh evidence and to discriminate among degrees of probable inference.

Instructions: Each problem consists of a short paragraph followed by several suggested conclusions. For the purpose of this test assume that everything in the short paragraph is true. The problem is to judge whether or not each of the proposed conclusions logically follows beyond reasonable doubt from the information given in the paragraph.

An example problem:

A Nottingham-based newspaper made a survey of the number of male and female drivers involved in all car accidents in the Nottingham area during a given period of time. They found that male drivers were involved in 1210 accidents while female drivers were involved in only 920 accidents.

- If the period studied is typical, more car accidents in the Nottingham area involve male drivers than female drivers.

  ○ FOLLOWS   ○ DOES NOT FOLLOW

- More men than women drive cars in the Nottingham area every day.

  ○ FOLLOWS   ○ DOES NOT FOLLOW

- Teenage boys are involved in car accidents more often than teenage girls in the Nottingham area.

  ○ FOLLOWS   ○ DOES NOT FOLLOW

Figure A.6 Interpretation (Watson–Glaser critical thinking appraisal).
Source: Watson, G. & Glaser, E. M. (1964). *Watson–Glaser Critical Thinking Appraisal.* San Diego: Hardcourt, Brace & World.

Recognition of Assumptions (Watson–Glaser Critical Thinking Appraisal)

Aim: To assess participants' ability to recognise unstated assumptions in given assertions or propositions.

Instructions: An assumption is something presupposed or taken for granted. Below are a number of statements. Each statement is followed by several proposed assumptions. You are to decide for each assumption whether a person, in making the given statement, is really making that assumption, *i.e.*, taking it for granted, justifiably or not.

An example problem:

Statement: "The discovery of additional ways of using atomic energy will, in the long run, prove a blessing to mankind".

Proposed Assumptions:

- Atomic energy can have numerous uses.
  - ○ MADE   ○ NOT MADE

- The discovery of additional uses for atomic energy will require large long-term investments of money.
  - ○ MADE   ○ NOT MADE

- The present uses of atomic energy are a curse to mankind.
  - ○ MADE   ○ NOT MADE

Figure A.7   Recognition of assumptions (Watson–Glaser critical thinking appraisal).
Source: Watson, G. & Glaser, E. M. (1964). *Watson–Glaser Critical Thinking Appraisal.* San Diego: Hardcourt, Brace & World.

Plausible Estimation

Aim: To assess participants' ability to construct plausible estimations.

Example problems:

Try to estimate reasonable answers to each of the following questions. Describe carefully at each stage any assumptions you make. Show, step by step, how you arrive at your estimate. You may assume that the population of the UK is 60 million.

- How many babies are born in the UK every minute?

- How many secondary school teachers are there in the UK?

- How many dentists are there in the UK?

Figure A.8 The estimation task.
Source: Swan, M. & Ridgway, J. (2010). *Plausible Estimation Tasks*, Field Tested Assessment Learning Guide.

Problem Solving

Aim: To assess participants' ability to solve problems.

Instructions: Each problem consists of a series of matchsticks arranged to form an arithmetic statement using Roman numerals. Your task is to correct the arithmetic statement by moving a single matchstick from one position in the statement to another.

Some example problems:

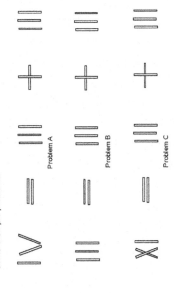

Figure A.9   Insight problem solving.
Source: Knoblich, G., Ohlsson, S., & Raney, G. E. (2001). An eye movement study of insight problem solving. *Memory & Cognition, 29,* 1000–1009.

Statistical Reasoning Task

Aim: To assess participants' ability to use statistical evidence to make decisions rather than personal experience.

Example Problem:

The admissions committee of the psychology department of a midlands university was considering which 10 of 30 applicants to admit to their PhD programme. The department keeps records on the performance of all its graduate students and relates this performance score to all kinds of background information about the students. There was a debate on the admissions committee about whether to admit a particular student from the University of Maynard. The student's marks on his degree course were marginal. Almost all students previously admitted to the department had higher marks. The student's references were very good, but none of the referees were personally known to any admissions faculty. One member of the admissions committee argued against admission, pointing out that department records show that students who graduate from small non-research universities like Maynard perform at a level substantially below the median of all students in the programme. This argument was countered by a committee member who noted that 2 years ago the university had admitted a student from Maynard who was now among the three highest ranked students in the department. What do you think the committee should do?

○ Definitely admit the student   ○ Probably admit the student   ○ Probably not admit the student

○ Definitely not admit the student

Figure A.10 The statistical reasoning task.
Source: Stanovich, K. E. & West, R. F. (1998). Individual differences in rational thought. *Journal of Experimental Psychology: General,* 127, 161–188.

Wason THOG Task

Aim: To assess participants' ability to analyse disjunctive statements.

Instructions:

In front of you are four designs: a black diamond, a white diamond, a black circle and a white circle.

You are to assume that I have written down one of the colours (black or white) and one of the shapes (diamond or circle). Now read the following rule carefully:

If, and only if, any of the designs includes either the colour I have written down, or the shape I have written down, but not both, then it is called a THOG.

I will tell you that the black diamond is a THOG. Each of the designs can now be classified into one of the following categories: (a) definitely is a THOG; (b) insufficient information to decide; (c) definitely is not a THOG. Your task is to classify the designs:

Black Diamond	○ THOG ○ insufficient information	○ not a THOG
White Diamond	○ THOG ○ insufficient information	○ not a THOG
Black Circle	○ THOG ○ insufficient information	○ not a THOG
White Circle	○ THOG ○ insufficient information	○ not a THOG

Figure A.11   The Wason THOG task.

Source: Wason, P. C. (1977). Self-contradictions. In P. Johnson-Laird & P. C. Wason (Eds.), *Thinking: Readings in Cognitive Science.* Cambridge: CUP.

Wason Selection Task

Aim: To assess participants' ability to analyse logical statements.

Instructions:

Four cards are placed on a table in front of you. Each card has a letter on one side and a number on the other.

You can see:

[D] [K] [3] [7]

Here is a conjectured rule about the cards:

> *"If a card has a D on one side, then it has a 3 on the other".*

Your task is to select all those cards, but only those cards, which you would need to turn over in order to find out whether the rule is true or false.

Figure A.12 The Wason selection task.

Source: Wason, P. C. (1968). Reasoning about a rule. *Quarterly Journal of Experimental Psychology, 20,* 273–281.

Appendix B

Abstract Conditional Inference Task

The task was based on that used by Evans et al. (1996).
Instructions:

This experiment is concerned with people's ability to reason logically with sentences in various forms. You will be presented with a total of 32 problems on the attached pages. In each case, you are given two statements together with a conclusion which may or may not follow from these statements.

Your task in each case is to decide whether or not the conclusion *necessarily* follows from the statements. A conclusion is necessary if it must be true, given that the statements are true.

Each problem concerns an imaginary letter–number pair and contains an initial statement or rule which determines which letters may be paired with which numbers. An example of a rule of similar form to those used would be:

- If the letter is B then the number is not 7.

In each case, you must assume that the rule holds and then combine it with the information given in the second statement. This will concern either the letter or the number of an imaginary pair, for example:

- The letter is Y.
- The number is not 4.

If the information concerns the letter the conclusion will concern the number and vice versa.

A full problem looks something like this:

If the letter is X then the number is 1.
The letter is X.
Conclusion: The number is 1.
    ○ YES
    ○ NO

If you think the conclusion necessarily follows please tick the YES box, otherwise tick the NO box.

Here are the 32 *problems* (*only the first* 16 *were used in the study reported in Chapter* 3). *Participants saw the problems in a random order*:

If the letter is A then the number is 3.
The letter is A.
Conclusion: The number is 3.
○ YES   ○ NO   (q.01)

If the letter is T then the number is 5.
The letter is not T.
Conclusion: The number is not 5.
○ YES   ○ NO   (q.02)

If the letter is F then the number is 8.
The number is 8.
Conclusion: The letter is F.
○ YES   ○ NO   (q.03)

If the letter is D then the number is 4.
The number is not 4.
Conclusion: The letter is not D.
○ YES   ○ NO   (q.04)

If the letter is G then the number is not 6.
The letter is G.
Conclusion: The number is not 6.
○ YES   ○ NO   (q.05)

If the letter is R then the number is not 1.
The letter is not R.
Conclusion: The number is 1.
○ YES   ○ NO   (q.06)

If the letter is K then the number is not 3.
The number is not 3.
Conclusion: The letter is K.
○ YES   ○ NO   (q.07)

If the letter is U then the number is not 9.
The number is 9.
Conclusion: The letter is not U.
○ YES   ○ NO   (q.08)

If the letter is not B then the number is 5.
The letter is not B.

Conclusion: The number is 5.
○ YES   ○ NO   (Q.09)

If the letter is not S then the number is 6.
The letter is S.
Conclusion: The number is not 6.
○ YES   ○ NO   (Q.10)

If the letter is not V then the number is 8.
The number is 8.
Conclusion: The letter is not V.
○ YES   ○ NO   (Q.11)

If the letter is not H then the number is 1.
The number is not 1.
Conclusion: The letter is H.
○ YES   ○ NO   (Q.12)

If the letter is not F then the number is not 3.
The letter is not F.
Conclusion: The number is not 3.
○ YES   ○ NO   (Q.13)

If the letter is not L then the number is not 9.
The letter is L.
Conclusion: The number is 9.
○ YES   ○ NO   (Q.14)

If the letter is not J then the number is not 8.
The number is not 8.
Conclusion: The letter is not J.
○ YES   ○ NO   (Q.15)

If the letter is not V then the number is not 7.
The number is 7.
Conclusion: The letter is V.
○ YES   ○ NO   (Q.16)

If the letter is D then the number is 2.
The letter is D.
Conclusion: The number is 2.
○ YES   ○ NO   (Q.17)

If the letter is Q then the number is 1.
The letter is K.
Conclusion: The number is not 1.
○ YES   ○ NO   (Q.18)

If the letter is M then the number is 4.
The number is 4.
Conclusion: The letter is M.
○ YES   ○ NO   (Q.19)

If the letter is V then the number is 5.
The number is 2.
Conclusion: The letter is not V.
○ YES   ○ NO   (Q.20)

If the letter is S then the number is not 8.
The letter is S.
Conclusion: The number is not 8.
○ YES   ○ NO   (Q.21)

If the letter is B then the number is not 3.
The letter is H.
Conclusion: The number is 3.
○ YES   ○ NO   (Q.22)

If the letter is J then the number is not 2.
The number is 7.
Conclusion: The letter is J.
○ YES   ○ NO   (Q.23)

If the letter is U then the number is not 7.
The number is 7.
Conclusion: The letter is not U.
○ YES   ○ NO   (Q.24)

If the letter is not E then the number is 2.
The letter is R.
Conclusion: The number is 2.
○ YES   ○ NO   (Q.25)

If the letter is not A then the number is 6.
The letter is A.
Conclusion: The number is not 6.
○ YES   ○ NO   (Q.26)

If the letter is not C then the number is 9.
The number is 9.
Conclusion: The letter is not C.
○ YES   ○ NO   (Q.27)

If the letter is not N then the number is 3.
The number is 5.

Conclusion: The letter is N.

○ YES   ○ NO   (Q.28)

If the letter is not A then the number is not 1.
The letter is N.
Conclusion: The number is not 1.

○ YES   ○ NO   (Q.29)

If the letter is not C then the number is not 2.
The letter is C.
Conclusion: The number is 2.

○ YES   ○ NO   (Q.30)

If the letter is not W then the number is not 8.
The number is 3.
Conclusion: The letter is not W.

○ YES   ○ NO   (Q.31)

If the letter is not K then the number is not 1.
The number is 1.
Conclusion: The letter is K.

○ YES   ○ NO   (Q.32)

Appendix C

Thematic Conditional Inference Task

The thematic conditional inference task used in the study reported in Chapter 3 had a similar structure to the abstract conditional inference task, except it had thematic content. It was based on the work of Evans *et al.* (2010). Participants saw the problems in a random order. The 16 problems are given below:

Premises:
If oil prices continue to rise then UK petrol prices will rise.
Oil prices continue to rise.
Conclusion:
UK petrol prices rise.

Premises:
If oil prices continue to rise then UK petrol prices will rise.
Oil prices do not continue to rise.
Conclusion:
UK petrol prices do not rise.

Premises:
If car ownership increases then traffic congestion will get worse.
Traffic congestion does not get worse.
Conclusion:
Car ownership does not increase.

Premises:
If car ownership increases then traffic congestion will get worse.
Traffic congestion gets worse.
Conclusion:
Car ownership increases.

Premises:
If more people use protective sun cream then cases of skin cancer will be reduced.
Cases of skin cancer are not reduced.

Conclusion:
More people do not use protective sun cream.

Premises:
If more people use protective sun cream then cases of skin cancer will be reduced.
Cases of skin cancer are reduced.
Conclusion:
More people use protective sun cream.

Premises:
If Sony release a PlayStation 4 then their company profits will rise.
Sony release a PlayStation 4.
Conclusion:
The company profits rise.

Premises: If Sony release a PlayStation 4 then their company profits will rise.
Sony do not release a Playstation 4.
Conclusion:
The company profits do not rise.

Premises:
If more new houses are built then the amount of homeless people will increase.
More new houses are built.
Conclusion:
The amount of homeless people increases.

Premises:
If more new houses are built then the amount of homeless people will increase.
More new houses are not built.
Conclusion:
The amount of homeless people does not increase.

Premises:
If third world debt is cancelled then world poverty will worsen.
World poverty does not worsen.
Conclusion:
Third world debt is not cancelled.

Premises:
If third world debt is cancelled then world poverty will worsen.
World poverty worsens.
Conclusion:
Third world debt is cancelled.

Premises:
If fast food is taxed then childhood obesity will increase.
Childhood obesity does not increase.
Conclusion:
Fast food is not taxed.

Premises:
If fast food is taxed then childhood obesity will increase.
Childhood obesity increases.
Conclusion:
Fast food is taxed.

Premises: If EU quarantine laws are strengthened then rabies will spread to the UK.
EU quarantine laws are strengthened.
Conclusion:
Rabies spreads to the UK.

Premises:
If EU quarantine laws are strengthened then rabies will spread to the UK.
EU quarantine laws are not strengthened.
Conclusion:
Rabies does not spread to the UK.

Appendix D

Abstract Syllogism Task

This section is concerned with people's ability to reason logically with sentences in various forms. You will be presented with a total of eight problems on the following pages.

In each case, you are given two statements together with a conclusion which may or may not follow from these statements. You must decide whether the conclusion follows logically from the premises or not. You must suppose that the premises are all true. Decide if the conclusion follows logically from the premises, assuming the premises are true, and tick your response.

Your task in each case is to decide whether or not the conclusion *necessarily* follows from the statements. *A conclusion is necessary if it must be true, given that the statements are true.*

A full problem looks something like this:

Premises:

All As are Bs.

All Bs are Cs.

Conclusion:

All As are Cs.

In each case you must assume that the rule holds and combine it with the information in the second statement in order to decide whether the conclusion necessarily follows.

If you think the conclusion necessarily follows then please select YES, otherwise select NO.

Premises:
All Ss are Es.
All Ds are Es.
Conclusion:
Alls Ds are Ss.

Premises:
All Fs are Gs.
All Hs are Fs.
Conclusion:
All Hs are Gs.

Premises:
All Ns are As.
No Ts are As.
Conclusion:
No Ts are Ns.

Premises:
All Ys are Ks.
No Ps are Ys.
Conclusion:
No Ps are Ks.

Premises:
All Ls are Rs.
All Cs are Rs.
Conclusion:
All Cs are Ls.

Premises:
All Ws are Is.
All Qs are Ws.
Conclusion:
All Qs are Is.

Premises:
All As are Ks.
No Ps are Ks.
Conclusion:
No Ps are As.

Premises:
All Bs are Vs.
No Ms are Bs.
Conclusion:
No Ms are Vs.

Appendix E

Thematic Syllogism Task

This task had a similar structure to the abstract syllogism task. The eight problems are given below:

Premises:
All lapitars wear clothes.
Podips wear clothes.
Conclusion:
Podips are lapitars.

Premises:
All ramadions taste delicious.
Gumthorps are ramadions.
Conclusion:
Gumthorps taste delicious.

Premises:
All selacians have sharp teeth.
Snorlups do not have sharp teeth.
Conclusion:
Snorlups are not selacians.

Premises:
All hudon are ferocious.
Wampets are not hudon.
Conclusion:
Wampets are not ferocious.

Premises:
All opprobines run on electricity.
Jamtops run on electricity.
Conclusion:
Jamtops are opprobines.

Premises:
All tumpers lay eggs.
Sampets are tumpers.

Conclusion:
Sampets lay eggs.

Premises:
All snapples run fast.
Alcomas do not run fast.

Conclusion:
Alcomas are not snapples.

Premises:
All argomelles are kind.
Magsums are not argomelles.
Conclusion:
Magsums are not kind.

Appendix F

Abstract Truth Table Task

Please read the following instructions carefully.

This section is concerned with people's ability to reason logically with sentences in various forms. You will be presented with a total of 32 problems on the following pages.

All problems relate to cards which have a capital letter on their left hand side and a single digit number on the right.

You will be given a rule together with a picture of a card to which the rule applies. Your task is to determine whether the card conforms to the rule, contradicts the rule, or is irrelevant to the rule.

A full problem looks something like this:

Rule: If the letter is A then the number is 1.

Card: | A 1 |

   ○   conforms to the rule

   ○   contradicts the rule

   ○   irrelevant to the rule

You should tick the option which you think best applies.

This section consists of this instructions page, and six pages of problems. Please work through the problems in order and make sure you do not miss any. Do not return to a problem once you have finished and moved on to another.

When you have finished this section, move straight on to the following section.

The questions begin here:

Rule: If the letter is B then the number is 3.
Card: ⏐B 3⏐
○ card conforms to the rule
○ card contradicts the rule
○ card irrelevant to the rule   (Q.1)

Rule: If the letter is B then the number is 3.
Card: ⏐B 7⏐
○ card conforms to the rule
○ card contradicts the rule
○ card irrelevant to the rule   (Q.2)

Rule: If the letter is B then the number is 3.
Card: ⏐N 3⏐
○ card conforms to the rule
○ card contradicts the rule
○ card irrelevant to the rule   (Q.3)

Rule: If the letter is B then the number is 3.
Card: ⏐G 4⏐
○ card conforms to the rule
○ card contradicts the rule
○ card irrelevant to the rule   (Q.4)

Rule: If the letter is S then the number is 7.
Card: ⏐S 7⏐
○ card conforms to the rule
○ card contradicts the rule
○ card irrelevant to the rule   (Q.5)

Rule: If the letter is S then the number is 7.
Card: ⏐S 2⏐
○ card conforms to the rule
○ card contradicts the rule
○ card irrelevant to the rule   (Q.6)

Rule: If the letter is S then the number is 7.
Card: ⏐H 7⏐
○ card conforms to the rule
○ card contradicts the rule
○ card irrelevant to the rule   (Q.7)

Rule: If the letter is S then the number is 7.
Card: ⏐W 1⏐
○ card conforms to the rule

○ card contradicts the rule
○ card irrelevant to the rule   (Q.8)

Rule: If the letter is G then the number is not 6.
Card: $\boxed{\text{G 9}}$
○ card conforms to the rule
○ card contradicts the rule
○ card irrelevant to the rule   (Q.9)

Rule: If the letter is G then the number is not 6.
Card: $\boxed{\text{G 6}}$
○ card conforms to the rule
○ card contradicts the rule
○ card irrelevant to the rule   (Q.10)

Rule: If the letter is G then the number is not 6.
Card: $\boxed{\text{T 4}}$
○ card conforms to the rule
○ card contradicts the rule
○ card irrelevant to the rule   (Q.11)

Rule: If the letter is G then the number is not 6.
Card: $\boxed{\text{A 6}}$
○ card conforms to the rule
○ card contradicts the rule
○ card irrelevant to the rule   (Q.12)

Rule: If the letter is V then the number is not 9.
Card: $\boxed{\text{V 8}}$
○ card conforms to the rule
○ card contradicts the rule
○ card irrelevant to the rule   (Q.13)

Rule: If the letter is V then the number is not 9.
Card: $\boxed{\text{V 9}}$
○ card conforms to the rule
○ card contradicts the rule
○ card irrelevant to the rule   (Q.14)

Rule: If the letter is V then the number is not 9.
Card: $\boxed{\text{D 4}}$
○ card conforms to the rule
○ card contradicts the rule
○ card irrelevant to the rule   (Q.15)

Rule: If the letter is V then the number is not 9.
Card: $\boxed{\text{E 9}}$
○ card conforms to the rule

○ card contradicts the rule
○ card irrelevant to the rule   (Q.16)

Rule: If the letter is not R then the number is 5.
Card: M 5
○ card conforms to the rule
○ card contradicts the rule
○ card irrelevant to the rule   (Q.17)

Rule: If the letter is not R then the number is 5.
Card: B 6
○ card conforms to the rule
○ card contradicts the rule
○ card irrelevant to the rule   (Q.18)

Rule: If the letter is not R then the number is 5.
Card: R 5
○ card conforms to the rule
○ card contradicts the rule
○ card irrelevant to the rule   (Q.19)

Rule: If the letter is not R then the number is 5.
Card: R 9
○ card conforms to the rule
○ card contradicts the rule
○ card irrelevant to the rule   (Q.20)

Rule: If the letter is not B then the number is 1.
Card: G 1
○ card conforms to the rule
○ card contradicts the rule
○ card irrelevant to the rule   (Q.21)

Rule: If the letter is not B then the number is 1.
Card: L 7
○ card conforms to the rule
○ card contradicts the rule
○ card irrelevant to the rule   (Q.22)

Rule: If the letter is not B then the number is 1.
Card: B 1
○ card conforms to the rule
○ card contradicts the rule
○ card irrelevant to the rule   (Q.23)

Rule: If the letter is not B then the number is 1.
Card: B 3
○ card conforms to the rule

○ card contradicts the rule
○ card irrelevant to the rule   (Q.24)

Rule: If the letter is not E then the number is not 1.
Card: ⬛ K 7
○ card conforms to the rule
○ card contradicts the rule
○ card irrelevant to the rule   (Q.25)

Rule: If the letter is not E then the number is not 1.
Card: ⬛ D 1
○ card conforms to the rule
○ card contradicts the rule
○ card irrelevant to the rule   (Q.26)

Rule: If the letter is not E then the number is not 1.
Card: ⬛ E 3
○ card conforms to the rule
○ card contradicts the rule
○ card irrelevant to the rule   (Q.27)

Rule: If the letter is not E then the number is not 1.
Card: ⬛ E 1
○ card conforms to the rule
○ card contradicts the rule
○ card irrelevant to the rule   (Q.28)

Rule: If the letter is not T then the number is not 6.
Card: ⬛ H 2
○ card conforms to the rule
○ card contradicts the rule
○ card irrelevant to the rule   (Q.29)

Rule: If the letter is not T then the number is not 6.
Card: ⬛ N 6
○ card conforms to the rule
○ card contradicts the rule
○ card irrelevant to the rule   (Q.30)

Rule: If the letter is not T then the number is not 6.
Card: ⬛ T 8
○ card conforms to the rule
○ card contradicts the rule
○ card irrelevant to the rule   (Q.31)

Rule: If the letter is not T then the number is not 6.
Card: $\boxed{\text{T 6}}$
◯ card conforms to the rule
◯ card contradicts the rule
◯ card irrelevant to the rule   (q.32)

Appendix G

The Cypriot Curriculum

G.1 Low-Intensity Curriculum: Year 1

- Revision (graph of a straight line, quadratic equation, graph of a parabola, Pythagoras's theorem)
- Progressions (arithmetic, geometric, solution and strategies of problems)
- Logarithms, Logarithmic and Exponential Functions (definition, properties, logarithmic equations, exponential equations, graphs of logarithmic and exponential functions)
- Consumer's Problems (proportion, percentage, partition, simple tax, income tax, VAT, compound interest)
- Trigonometry (revision, trigonometric number of any angle, sine rule, cosine rule, area of a triangle, solution of triangle)
- Geometry (revision of flat shapes, similar shapes, regular polygons, circle, area of composite shapes)

G.2 Low-Intensity Curriculum: Year 2

- Revision
- Statistics (basic concepts, presentation of statistical data, characteristic values of a distribution)
- Combinations (basic counting principle, permutations, conditions, combinations)
- Probabilities
- Stereometry (polyhedral, solid shapes formed by rotation of flat shapes)
- Consumer's Problems (revision, motion problems)

G.3   High-Intensity Curriculum: Year 1

- Revision
- Absolute Value of Real Numbers (the group of real numbers, absolute value of a real number, types of intervals, equations and inequalities with absolute values)
- Functions (revision, function domain given in the form of a formula, equality and operations between functions, combination of functions, one-to-one function)
- Limit of a Function (the concept of a limit tending to plus or minus infinity, properties, allowed and not allowed actions between the symbols of plus and minus infinity and real numbers, limit of a function)
- Complex Numbers (imaginary numbers, definition of complex numbers, operations, properties)
- Perfect Induction (method of perfect induction)
- Sequences (the concept of a sequence, ways of defining a sequence, monotonic sequences, sequence limit, the symbolism)
- Progressions (arithmetic, geometric, formula of the nth term, properties, sum of n consecutive terms, sum of infinite terms of a geometric progression)
- Exponential Function — Logarithmic Function (introduction, the exponential function, exponential equations and inequalities, law of exponential change, concept of logarithm, properties, logarithmic function, logarithmic equations)
- Continuous Function (concept and definition of continuity, properties of continuous functions)
- Function Derivative (differentiability and continuity, derivatives of basic functions, differentiation rules, trigonometric function derivatives, derivative of complex function, implicit function derivative, higher order derivatives, inverse function derivative, derivative applications, tangent of curve)
- Polynomials (polynomials of a variable, fraction analysis into a sum of simple fractions)
- Determinants (vectors on a plane, vector operations, Cartesian coordinates of a point and vector, internal product of two vectors)
- Equation of a Line
- Sine rule, Cosine rule, Area of a Triangle
- Trigonometric Numbers of the Sum and Difference of Two Arcs

- Transformation of a Sum of Trigonometric Numbers to a Product and Vice Versa
- Trigonometric Equations
- Quadrilaterals Inscribed in Circle
- Regular Polygons
- Circle Measurement
- Concept and Basic Locus of Points
- Analytic and Combinatory Method
- Plane in Space
- Relative Position of Straight Line and Plane in Space
- Vertical Straight Lines on a Plane
- Parallel Lines and Planes in Space
- Direct Projection of a Line Towards a Plane and Line Slope
- Angle Between Two Planes
- Prisms and Pyramids
- Solid Shapes Formed by Rotation

G.4 High-Intensity Curriculum: Year 2

- Functions (revision, functions which are defined parametrically)
- Function Derivative (revision, derivatives of functions which are defined parametrically, derivative applications, differential of a function)
- Graphs (introduction, calculus' mean value theorem, monotonic function, local extremities, concave/convex function and points of inflection, asymptote lines of function diagram, graphs of functions, maximum and minimum problems)
- Inverse Trigonometric Functions (revision, inverse trigonometric functions)
- Indefinite Integral (definition, formulae of basic integrals, properties, integration by factors, integration of rational functions)
- Defined Integral (definition and calculation of a defined integral, properties, applications)
- Combinations (basic counting principle and applications, properties of combinations n choose k)
- Probability Theory (probabilities)
- Matrices (matrices)
- Locus of Points
- Conic Intersections (circle, conic intersections in general, parabola, ellipse, isosceles hyperbola)

Appendix H

Proof Comprehension

H.1 The Proof

Note: An *abundant* number is a positive integer n whose divisors add up to more than $2n$.

Theorem
The product of two distinct primes is not abundant.

Proof
- **(L1)** Let $n = p_1 p_2$, where p_1 and p_2 are distinct primes.
- **(L2)** Assume that $2 \leq p_1$ and $3 \leq p_2$.
- **(L3)** The divisors of n are $1, p_1, p_2$ and $p_1 p_2$.
- **(L4)** Note that $\frac{p_1+1}{p_1-1}$ is a decreasing function of p_1.
- **(L5)** So $\max\{\frac{p_1+1}{p_1-1}\} = \frac{2+1}{2-1} = 3$.
- **(L6)** Hence, $\frac{p_1+1}{p_1-1} \leq p_2$.
- **(L7)** So $p_1 + 1 \leq p_1 p_2 - p_2$.
- **(L8)** So $p_1 + 1 + p_2 \leq p_1 p_2$.
- **(L9)** So $1 + p_1 + p_2 + p_1 p_2 \leq 2p_1 p_2$. $\qquad\qquad$ $\square$

H.2 Proof Comprehension Items

(1) What are the three smallest abundant numbers?

- ○ $12, 18, 20$.
- ○ $6, 12, 18$.
- ○ $12, 24, 36$.

(2) How do we know that the divisors of n are $1, p_1, p_2$ and $p_1 p_2$?

- ○ We don't — there might be other divisors too, but these are the ones we need for this proof.
- ○ Because p_1 and p_2 are distinct primes.

○ Because $2 \leq p_1$ and $3 \leq p_2$ and in the equality case we would have $p_1 = 2$, $p_2 = 3$ and $p_1 p_2 = 6$.

(3) How do we know that $\frac{p_1+1}{p_1-1}$ is a decreasing function of p_1?

○ Because p_1 is a prime number greater than or equal to 2.
○ Because $p_1 + 1 > p_1 - 1$ for every prime number p_1.
○ Because $\frac{p_1+1}{p_1-1} = \frac{p_1-1+2}{p_1-1} = 1 + \frac{2}{p_1-1}$, which decreases as p_1 increases.

(4) Why is it valid to assume that $2 \leq p_1$ and $3 \leq p_2$?

○ Because we need this to prove that $\frac{p_1+1}{p_1-1} \leq p_2$.
○ Because all primes are greater than or equal to 2 and the two primes p_1 and p_2 are distinct.
○ This is not a question of validity — we can assume anything we like at the start of a proof.

(5) Which of the following best describes the logical relationship between lines (L2), (L5) and (L6)?

○ (L6) logically depends on statements made in lines (L2) and (L5).
○ (L6) logically depends on statements made in line (L5) but not in (L2).
○ The lines are logically independent.

(6) Which of the following best describes the logical relation between lines (L2) and (L4)?

○ (L2) logically depends on statements made in line (L4).
○ The lines are logically independent.
○ (L4) logically depends on statements made in line (L2).

(7) Which of the following summaries best captures the ideas of the proof?

○ The proof uses the fact that the divisors of n are $1, p_1, p_2$ and $p_1 p_2$ to prove that $\max\{\frac{p_1+1}{p_1-1}\} = 3$.
  It then assumes that $\max\{\frac{p_1+1}{p_1-1}\} \leq p_2$ and shows this implies that $1 + p_1 + p_2 + p_1 p_2 \leq 2 p_1 p_2$.
○ The proof uses the assumption that the divisors p_1 and p_2 are prime and distinct to establish an inequality relationship between $\frac{p_1+1}{p_1-1}$ and p_2.
  This inequality is then manipulated to show that the sum of the divisors of n is less than or equal to $2n$.
○ The proof establishes that $\frac{p_1+1}{p_1-1}$ is a decreasing function of p_1, in order to conclude that it is eventually less than or equal to p_2.
  It then uses the fact that $p_1 + 1 \leq p_1 p_2 - p_2$ to reach the conclusion.

(8) Can we conclude from this proof that $n = p_1 p_2$ is not abundant, if $3 \leq p_1 \leq p_2$ and p_2 is prime but p_1 is not?

○ Yes, because in such a case $\frac{p_1 + 1}{p_1 - 1}$ is still a decreasing function of p_1.

○ Yes, because in such a case it is still true that $2 \leq p_1$ and $3 \leq p_2$.

○ No, because in such a case the divisors of n are not $1, p_1, p_2$ and $p_1 p_2$.

(9) Using the logic of the proof, which best explains why the divisors of 39 add up to a number less than or equal to 78?

○ $39 = 3 \times 13$ and $1 + 3 + 13 + 39 = 56 \leq 78$.

○ $39 = 3 \times 13$, which is the product of two distinct primes so its divisors are $1, 3, 13$ and 39. $(3 + 1)/(3 - 1) \leq 13$ so rearranging and adding 39 to both sides gives $1 + 3 + 13 + 39 \leq 2 \times 3 \times 13 \leq 78$.

○ $39 = 3 \times 13$ and $(3 + 1)(3 - 1)$ is a decreasing function of 3, so $1 + 3 + 13 + 39 \leq 2 \times 3 \times 13 = 78$.

(10) Could lines (L4) and (L5) be replaced with a different subproof that $\max\{\frac{p_1 + 1}{p_1 - 1}\} \leq 3$?

○ No, because line (L4) is needed to make the link to p_1 in $n = p_1 p_2$.

○ Yes, because all we need for the rest of the proof is that $\frac{p_1 + 1}{p_1 - 1} \leq p_2$.

○ No, because p_1 might be equal to 2 so it is important that this maximum is less than or equal to 3.

Bibliography

Adkins, M., & Noyes, A. (2016). Reassessing the economic value of advanced level mathematics. *British Educational Research Journal*, *42*, 93–116.

Alcock, L., Bailey, T., Inglis, M., & Docherty, P. (2014). The ability to reject invalid logical inferences predicts proof comprehension and mathematics performance. In T. Fukawa-Connolly, G. Karakok, K. Keene, & M. Zandieh (Eds.), *Proceedings of the 17th Annual Conference on Research in Undergraduate Mathematics Education* (pp. 376–383). Denver, CO: MAA.

Alcock, L., Hodds, M., Roy, S., & Inglis, M. (2015). Investigating and improving undergraduate proof comprehension. *Notices of the American Mathematical Society*, *62*, 742–752.

Argyropoulos, E., Vlamos, P., Katsoulis, G., Markatis, S., & Sideris, P. (2010). *Euclidean Geometry General Lyceum A and B*. Athens, Greece: Organization Textbook Publishing.

Armstrong, M. A. (1983). *Basic Topology*. New York: Springer.

Assessment and Qualifications Alliance. (2015). *General Certificate of Education Mathematics Specification*. Retrieved 19th May 2015, from http://bit.ly/1DOVVV2.

Attridge, N., Doritou, M., & Inglis, M. (2015). The development of reasoning skills during compulsory 16 to 18 mathematics education. *Research in Mathematics Education*, *17*, 20–37.

Attridge, N., & Inglis, M. (2013). Advanced mathematical study and the development of conditional reasoning skills. *PLOS ONE*, *8*, e69399.

Attridge, N., & Inglis, M. (2014). Intelligence and negation biases on the conditional inference task. *Thinking and Reasoning*, *20*, 454–471.

Ayton, P., & Hardman, D. (1997). Are two rationalities better than one? *Current Psychology of Cognition*, *16*, 39–51.

Bacon, F. (1625). Of studies. In F. Bacon (Ed.), *The Essayes or Counsels, Civill and Morall, of Francis Lo Verulam, Viscount St. Alban*. EP Dutton & Co, London.

Bagley, W. C. (1905). *The Educative Process*. London: Macmillan.

Bennett, A., Cuthbert, D., Fawcett, H., Hartung, M., Havighurst, R., Jablonower, J. & Thayer, V. T. (1938). *Mathematics in General Education: A Report of the Committee on the Function of Mathematics in General Education for the Commission on Secondary School Curriculum*. New York: Appleton.

Broyler, C. R., Thorndike, E. L., & Woodyard, E. (1927). A second study of mental discipline in high school studies. *Journal of Educational Psychology, 18*, 377–404.

Bruine de Bruin, W., Parker, A. M., & Fischoff, B. (2007). Individual differences in adult decision-making competence. *Journal of Personality and Social Psychology, 92*, 938–956.

Bursill-Hall, P. (2002). *Why do We Study Geometry? Answers through the Ages*. (Part of the Opening Festivities of the Faulkes Institute for Geometry, Cambridge, UK).

Cacioppo, J. T., Petty, R. E., Feinstein, J. A., & Jarvis, W. B. G. (1996). Dispositional differences in cognitive motivation: The life and times of individuals varying in need for cognition. *Psychological Bulletin, 119*, 197–253.

Cheng, P. W., Holyoak, K. J., Nisbett, R. E., & Oliver, L. M. (1986). Pragmatic versus syntactic approaches to training deductive reasoning. *Cognitive Psychology, 18*, 293–328.

Cockcroft, W. H. (1982). *Mathematics Counts*. London: HMSO.

Cohen, L. J. (1981). Can human irrationality be experimentally demonstrated? *Behavioural and Brain Sciences, 4*, 317–370.

Deary, I. J. (2001). *Intelligence: A Very Short Introduction*. Oxford: Oxford University Press.

Dolton, P. J., & Vignoles, A. (2002). The return on post-compulsory school mathematics study. *Economica, 69*, 113–141.

Durand-Guerrier, V. (2003). Which notion of implication is the right one? From logical considerations to a didactic perspective. *Educational Studies in Mathematics, 53*, 5–34.

Evans, J. St. B. T. (1977). Linguistic factors in reasoning. *Quarterly Journal of Experimental Psychology, 29*, 297–306.

Evans, J. St. B. T. (2006). The heuristic-analytic theory of reasoning: Extension and evaluation. *Psychonomic Bulletin and Review, 13*, 378–395.

Evans, J. St. B. T., & Ball, L. J. (2010). Do people reason on the selection task? A new look at the data of Ball *et al.* (2003). *Quarterly Journal of Experimental Psychology, 63*, 434–441.

Evans, J. St. B. T., Barston, J. L., & Pollard, P. (1983). On the conflict between logic and belief in syllogistic reasoning. *Memory & Cognition, 11*, 295–306.

Evans, J. St. B. T., Clibbens, J., & Rood, B. (1995). Bias in conditional inference: Implications for mental models and mental logic. *Quarterly Journal of Experimental Psychology, 48A*(3), 644–670.

Evans, J. St. B. T., Clibbens, J., & Rood, B. (1996). The role of implicit and explicit negation in conditional reasoning bias. *Journal of Memory and Language, 35*, 392–409.

Evans, J. St. B. T., & Handley, S. J. (1999). The role of negation in conditional inference. *Quarterly Journal of Experimental Psychology, 52A*, 739–769.

Evans, J. St. B. T., Handley, S. J., Neilens, H., & Over, D. (2010). The influence of cognitive ability and instructional set on causal conditional inference. *Quarterly Journal of Experimental Psychology, 63*, 892–909.

Evans, J. St. B. T., Handley, S. J., Neilens, H., & Over, D. E. (2007). Thinking about conditionals: A study of individual differences. *Memory & Cognition, 35*, 1772–1784.

Evans, J. St. B. T., & Lynch, J. S. (1973). Matching bias in the selection task. *British Journal of Psychology, 64*, 391–397.

Evans, J. St. B. T., & Over, D. E. (2004). *If.* Oxford: OUP.

Evans, J. St. B. T., & Stanovich, K. E. (2013). Dual-process theories of higher cognition: Advancing the debate. *Perspectives on Psychological Science, 8*, 223–241.

Fitch, J. G. (1883). *Lectures on Teaching.* Cambridge: Cambridge University Press.

Flynn, J. R. (2009). *What is Intelligence? Beyond the Flynn effect.* Cambridge: Cambridge University Press.

Fong, G. T., Krantz, D. H., & Nisbett, R. E. (1986). The effects of statistical training on thinking about everyday problems. *Cognitive Psychology, 18*, 253–292.

Frederick, S. (2005). Cognitive reflection and decision making. *Journal of Economic Perspectives*, *19*, 25–42.

Gates, A. I. (1949). Edward L Thorndike 1874–1949. *Psychological Review*, *56*, 241–243.

Gonzálex, G., & Herbst, P. G. (2006). Competing arguments for the geometry course: Why were American high school students supposed to study geometry in the twentieth century? *International Journal for the History of Mathematics Education*, *1*, 1.

Gross, J. (2008). *The long term costs of numeracy difficulties*. Every Child a Chance Trust and KPMG.

Heim, A. W. (1968). *AH5 Group Test of Intelligence*. London: National Foundation for Educational Research.

Hendrickson, G., & Schroeder, W. H. (1941). Transfer of training in learning to hit a submerged target. *Journal of Educational Psychology*, *32*, 205–213.

Hodgen, J., Pepper, D., Sturman, L., & Ruddock, G. (2010). *Is the UK an Outlier? An International Comparison of Upper Secondary Mathematics Education*. London: Nuffield Foundation.

Howson, G. (1982). *A History of Mathematics Education in England*. Cambridge: Cambridge University Press.

Hoyles, C., & Küchemann, D. (2002). Students' understanding of logical implication. *Educational Studies in Mathematics*, *51*, 193–223.

Inglis, M., & Alcock, L. (2012). Expert and novice approaches to reading mathematical proofs. *Journal for Research in Mathematics Education*, *43*, 358–390.

Inglis, M., Croft, T., & Matthews, J. (2011). *Graduates' Views on the Undergraduate Mathematics Curriculum*. Birmingham: National HE STEM Programme.

Inhelder, B., & Piaget, J. (1958). *The Growth of Logical Thinking from Childhood to Adolescence: An Essay on the Construction of Formal Operational Structures*. New York: Basic Books.

Johnson-Laird, P. N., & Byrne, R. M. J. (2002). Conditionals: A theory of meaning, pragmatics and inference. *Psychological Review*, *109*, 646–678.

Joint Council for Qualifications. (2014). *GCSE and Entry Level Certificate Results Summer 2014*. London.

Judd, C. H. (1908). The relation of special training and general intelligence. *Educational Review*, *36*, 28–42.

Judge, T. A., Higgins, C. A., Thoresen, C. J., & Barrick, M. R. (1999). The big five personality traits, general mental ability, and career success across the life span. *Personnel Psychology, 52,* 621–652.

Kahneman, D. (1981). Who shall be the arbiter of our intuitions? *Behavioral and Brain Sciences, 4,* 339–340.

Kahneman, D. (2011). *Thinking, Fast and Slow.* New York: Farrar, Straus and Giroux.

Klauer, K. C., Stahl, C., & Erdfelder, E. (2007). The abstract selection task: New data and an almost comprehensive model. *Journal of Experimental Psychology: Learning, Memory and Cognition, 33,* 680–703.

Krantz, J. H., & Dalal, R. (2000). Validity of web-based psychological research. In M. H. Birnbaum (Ed.), *Psychological Experiments on the Internet* (pp. 35–60). San Diego: Academic Press.

Lamont, A., & Maton, K. (2008). Choosing music: Exploratory studies into the low uptake of music GCSE. *British Journal of Music Education, 25,* 267–282.

Larrick, R. P., Nisbett, R. E., & Morgan, J. N. (1993). Who uses the cost-benefit rules of choice? Implications for the normative status of microeconomic threory. *Organizational Behaviour and Human Decision Processes, 56,* 331–347.

Lave, J. (1988). *Cognition in Practice: Mind, Mathematics, and Culture in Everyday Life.* Cambridge: CUP.

Lehman, D. R., Lempert, R. O., & Nisbett, R. E. (1988). The effects of graduate training on reasoning. *American Psychologist, 43,* 431–442.

Lehman, D. R., & Nisbett, R. E. (1990). A longitudinal study of the effects of undergraduate training on reasoning. *Developmental Psychology, 26,* 952–960.

Lobato, J. (2006). Alternative perspectives on the transfer of learning: History, issues, and challenges for future research. *Journal of the Learning Sciences, 15,* 431–449.

Locke, J. (1706/1971). *Conduct of the Understanding.* New York: Burt Franklin.

Locke, J. (1854). *An Essay Concerning Human Understanding and a Treatise on the Conduct of the Understanding.* Philadelphia, PA: Hayes & Zell Publishers.

Lopes, L. L. (1991). The rhetoric of irrationality. *Theory and Psychology, 1,* 65–82.

Manktelow, K. (2012). *Thinking and Reasoning: An Introduction to the Psychology of Reason, Judgment and Decision Making.* Hove, UK: Psychology Press.

Mann, L. (1979). *On the Trail of Process: A Historical Perspective on Cognitive Processes and Their Training.* New York: Grune & Stratton.

Martin, U., & Pease, A. (2013). *What does mathoverflow tell us about the production of mathematics?* Retrieved from http://arxiv.org/abs/1305.0904.

Maxwell, E. A. (1959). *Fallacies in Mathematics.* Cambridge University Press.

Mitchell, D. (1962). *An Introduction to Logic.* London: Hutchinson.

Monaghan, J. (1991). Problems with the language of limits. *For the Learning of Mathematics, 11*(3), 20–24.

Monroe, P. (1909). *A Text-book in the History of Education.* London: Macmillan.

Morely, N. J., Evans, J. St. B. T., & Handley, S. J. (2004). Belief bias and figural bias in syllogistic reasoning. *Quarterly Journal of Experimental Psychology, 57*, 666–692.

Morris, M. W., & Nisbett, R. E. (1992). Tools of the trade: Deductive reasoning schemas taught in psychology and philosophy. In R. E. Nisbett (Ed.), *Rules for reasoning* (pp. 228–256). Hillsdale, N.J.: Erlbaum.

Mouzakitis, A. (2006). A comparative analysis of Italian and Greek Euclidean geometry textbooks: A case study. *Philosophy of Mathematics Education Journal, 19*.

Newell, A. (1980). One last word. In D. Tuma & F. Reif (Eds.), *Problem Solving and Education.* New Jersey: Erlbaum.

Nisbett, R. E. (2009). Can reasoning be taught? In E. Callan, T. Grotzer, J. Kagan, R. E. Nisbett, D. N. Perkins, & L. S. Shulman (Eds.), *Education and a Civic Society: Teaching Evidence-based Decision Making.* Cambridge, MA: American Academy of Arts & Sciences.

Nisbett, R. E., Fong, G. T., Lehman, D. R., & Cheng, P. W. (1987). Teaching reasoning. *Science, 238*, 625–631.

Oaksford, M., & Chater, N. (1994). A rational analysis of the selection task as optimal data selection. *Psychological Review, 101*, 608–631.

Oaksford, M., & Chater, N. (2007). *Bayesian Rationality: The Probabilistic Approach to Human Reasoning.* Oxford: Oxford University Press.

OCR. (2003). *Examiner's report for MEI Structured Mathematics Advanced Level, Decision and Discrete Mathematics 2.*

Osgood, W. F. (1903). A Jordan curve of positive area. *Transactions of the American Mathematical Society*, *4*, 107–112.

Parker, A. M., Bruine de Bruin, W., & Fischoff, B. (2015). Negative decision outcomes are more common with lower decision-making competence: An item-level analysis of the Decision Outcomes Inventory (DOI). *Frontiers in Psychology*, *6*, 363.

Parsons, S., & Bynner, J. (2005). *Does Numeracy Matter More?* London: NRDC.

Plato. (375BC/2003). *The Republic* (D. Lee, Ed.). London: Penguin.

Reid, D., & Inglis, M. (2005). Talking about logic. *For the Learning of Mathematics*, *25*(2), 24–25.

Richardson, E. P. (1979). Benjamin Franklin's chessmen. *American Art Journal*, *11*, 58–61.

Rota, G.-C. (1997). The phenomenology of mathematical beauty. *Synthèse*, *111*, 171–182.

Sá, W. C., West, R. F., & Stanovich, K. E. (1999). The domain specificity and generality of belief bias: Searching for a generalizable critical thinking skill. *Journal of Educational Psychology*, *91*, 497–510.

Sagan, H. (1994). *Space-filling Curves*. New York: Springer-Verlag.

Schroyens, W., Schaeken, W., & d'Ydewalle, G. (2001). The processing on negations in conditional reasoning: A meta-analytical case study in mental model and/or mental logic theory. *Thinking and Reasoning*, *7*, 121–172.

Selden, A., & Selden, J. (2003). Validations of proofs considered as texts: Can undergraduates tell whether an argument proves a theorem? *Journal for Research in Mathematics Education*, *34*, 4–36.

Smith, A. (2004). *Making Mathematics Count: The Report of Professor Adrian Smith's Inquiry into post-14 Mathematics Education*. London: The Stationery Office.

Stanic, G. M. A. (1986). The growing crisis in mathematics education in the early twentieth century. *Journal for Research in Mathematics Education*, *17*, 190–205.

Stanovich, K. E. (1999). *Who is Rational? Studies of Individual Differences in Reasoning*. Mahwah, NJ: Lawrence Erlbaum.

Stanovich, K. E. (2004). *The Robot's Rebellion: Finding Meaning in the Age of Darwin*. Chicago: Chicago University Press.

Stanovich, K. E. (2009). *What Intelligence Tests Miss: The Psychology of Rational Thought*. New Haven, CT: Yale University Press.

Stanovich, K. E., & Cunningham, A. E. (1992). Studying the consequences of literacy within a literate society: The cognitive correlates of print exposure. *Memory & Cognition, 20,* 51–68.

Stanovich, K. E., & West, R. F. (1998). Individual differences in rational thought. *Journal of Experimental Psychology: General, 127,* 161–188.

Stanovich, K. E., & West, R. F. (2000). Individual differences in reasoning: Implications for the rationality debate? *Behavioural and Brain Sciences, 23,* 645–726.

Tall, D. O., & Vinner, S. (1981). Concept image and concept definition in mathematics with particular reference to limits and continuity. *Educational Studies in Mathematics, 12,* 151–169.

Thorndike, E. L. (1906). *The Principles of Teaching: Based on Psychology.* New York: A G Seiler.

Thorndike, E. L. (1922). Instruments for measuring the disciplinary value of studies. *Journal of Educational Research, 5,* 269–279.

Thorndike, E. L. (1924). Mental discipline in high school studies. *Journal of Educational Psychology, 15,* 1–22.

Thorndike, E. L., & Woodworth, R. S. (1901a). The influence of improvement in one mental function upon the efficiency of other functions (I). *Psychological Review, 8,* 247–261.

Thorndike, E. L., & Woodworth, R. S. (1901b). The influence of improvement in one mental function upon the efficiency of other functions: III Functions involving attention, observation and discrimination. *Psychological Review, 8,* 553–564.

Thorndike, E. L., & Woodworth, R. S. (1901c). The influence of improvement in one mental function upon the efficiency of other functions: II The estimation of magnitudes. *Psychological Review, 8,* 384–395.

Todd, C. S. (2008). Unmasking the truth beneath the beauty: Why the supposed aesthetic judgements made in science may not be aesthetic at all. *International Studies in the Philosophy of Science, 11,* 61–79.

Toplak, M. E., West, R. F., & Stanovich, K. E. (2011). The Cognitive Reflection Test as a predictor of performance on heuristics and biases tasks. *Memory & Cognition, 39,* 1275–1289.

Tversky, A., & Kahneman, D. (1981). The framing of decisions and the psychology of choice. *Science, 211,* 453–458.

Vorderman, C., Porkess, R., Budd, C., Dunne, R., Rahman-Hart, P., Colmez, C., & Lee, S. (2011). *A World Class Mathematics Education for all our Young People.* London: Conservative Party.

Wainwright, E., Attridge, N., Wainwright, D., Alcock, L., & Inglis, M. (2016). Does studying mathematics improve thinking skills? Views of UK policymakers. *Manuscript under review*.

Wason, P. C. (1960). On the failure to eliminate hypotheses in a conceptual task. *Quarterly Journal of Experimental Psychology, 12*, 129–140.

Wason, P. C. (1966). Reasoning. In B. Foss (Ed.), *New Horizons in Psychology* (pp. 135–151). Harmondsworth: Penguin Books.

Wason, P. C., & Johnson-Laird, P. N. (1972). *Psychology of Reasoning*. London: B. T. Batsford.

Wason, P. C., & Shapiro, D. (1971). Natural and contrived experience in a reasoning problem. *Quarterly Journal of Experimental Psychology, 23*, 63–71.

Watts, I. (1801). *The Improvement of the Mind, or a Supplement to the Art of Logic*. Google Play.

White, J. (2000). Should mathematics be compulsory for all? In S. Bramall & J. White (Eds.), *Why learn maths?* (pp. 71–84). London: Institute of Education, University of London.

Whyburn, G. (1942). What is a curve? *American Mathematical Monthly, 49*, 493–497.

Index

Printed in the United States
By Bookmasters